Sun Sight Navigation

Celestial for Sailors

By ARTHUR A. BIRNEY
Design and Illustrations by E. JAMES WHITE

Cornell Maritime Press
Centreville, Maryland

Library of Congress Cataloging in Publication Data

Birney, Arthur A., 1927-
  Sun sight navigation.

  1. Nautical astronomy.   2. Navigation.   I. Title.
VK555.B64   1984   623.89   83-46034
ISBN 0-87033-318-6

Manufactured in the United States of America, First Edition

Table of Contents

Part Two: Sun Lines Any Time

PART ONE

NOON SIGHT NAVIGATION

Introduction

Part I of this little book, Noon Sight Navigation, offers an extremely simple, accurate, and fast method of establishing one's position at sea. It can be read easily in an hour or two, and any sailor of average competence should be able to use the method herein described without difficulty once he has read it.

Noon sight navigation is not a new concept. It has been used and proven by many mariners since Captain Cook began his famous voyages in 1768; it is practical and it works. The essentials are presented here in a fresh and easily understood series of steps which lead to the quick determination of a position.

This booklet is not intended as a course in basic piloting or seamanship, nor is it intended to be the ultimate authority on celestial navigation. It is intended for the reasonably experienced skipper who for various reasons has not learned complete celestial navigation but may from time to time find himself offshore and wondering where in the world he is. In the event of the navigator's sudden illness or if he should jump ship in Tahiti, the procedures here described will bring both ship and crew home safely. And, really, that is all that one can ask.

The obvious merits of this system are its simplicity and speed. Its major drawback is that it can only be used at one time during the day—at local apparent noon (LAN), when the sun is at its highest. If the sun or the horizon should be obscured at that particular moment, the opportunity for a noon sight is lost until the next day.

Most experienced navigators use the noon sight to establish latitude but few seem to realize that longitude can be obtained as well. Beyond the requirements of the usual latitude sight, it is only necessary to take the time and to follow the procedure detailed to obtain longitude.

Part II, Sun Lines Any Time, shows how to obtain lines of position from the sun, cross them, and thus obtain a fix at any time when solar observations are possible. It requires a few additional steps over the noon sight but the ability to use this procedure at any time when the sun is visible makes it worth the extra effort. With the use of the simplified Work Form supplied herein, a line of position may be quickly and simply obtained using only the simplest addition and subtraction.

In this age of electronic wizardry when black boxes quickly and accurately will give positions at the press of a few buttons, one may well ask if it is worthwhile to learn celestial navigation. Experienced blue water sailors will almost unanimously agree that it is. Salt air and moisture are the implacable foes of electronic circuits, and it is rare indeed to finish an open water passage with all of the electronics still working. As a case in point, when our boat finished the 1982 Bermuda Race, all of the following had failed: loran, wind speed, log (both speed and distance), and the depth sounder. The only electric items still functioning were the lights and the radio.

Even though we know that the earth spins on its axis daily while traveling around the sun annually, for our purposes, as practical navigators, we shall consider the sun as circling the earth, as this is what appears to happen. Thus, the sun seems to rise in the east each morning, to travel across the sky, and to set in the west each evening.

Chapter One: Basics

Equipment

To navigate successfully one needs certain basic equipment. This includes a proper ship's compass, parallel rules, dividers, a sextant, an accurate timepiece, charts, plotting sheets, a current *Nautical Almanac* and Sight Reduction Tables (based on *Sight Reduction Tables for Air Navigation* (Pub. No. 249).* A radio receiver capable of picking up time signals is also highly desirable as it will permit ship's time to be maintained to the second. The importance of extremely accurate time will be seen later.

Dead Reckoning

Dead reckoning (DR) is the art of establishing a position at sea by drawing a course line on a chart from a known location in the direction of one's compass course and marking off on this line the distance traveled over a given period of time. Distance is determined by multiplying estimated speed by the time traveled. The term "dead reckoning" is probably derived as a contraction of the expression "deduced" or "ded" reckoning. Accurate dead reckoning is the single most important element in successful navigation. Columbus discovered America with nothing more than DR, and with it he returned to his landfalls in the New World on the three voyages following his great discovery. When all else fails for the navigator (and you may rest assured that from time to time all else will fail), he will have his DR to fall back on.

If one were to sail from Sandy Hook at New York Harbor at 10:00 A.M. on a true course of 90° maintaining a speed of 6 knots, he would be exactly 18 miles due east of Sandy Hoo at the end of three hours; 24 miles at the end of 4 hours, and so forth. This is dead reckoning in its simplest form.

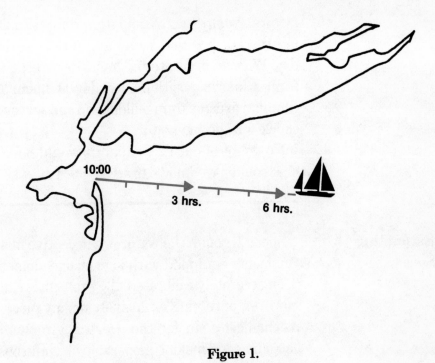

Figure 1.

Setting up the DR and making sure that proper entries are regularly made are primarily the navigator's, but ultimately the skipper's, responsibilities. Where there are separate watches, as in ocean racing, the navigator should explain precisely what is needed to the watch captains and at frequent regular intervals plot the information entered in the rough log by the watch as a DR position on the chart or plotting sheet. An entry should be made in the log at least every half hour indicating the time of the entry and the course and speed averaged during the previous half hour. In addition, of course, the exact time of any change of course should be noted.

Errors creep into the DR position as a result of currents, leeway, inaccurate estimates of speed, storms which render estimates of speed and course difficult and various other factors. *Noon Sight Navigation* offers a method of correcting the DR position by an accurate celestial fix once each day. With an accurate fix you can correct the DR and start from a known position each day at noon. This prevents a cumulative error from building up and serves as an invaluable check for the navigator.

No doubt about it, Columbus would have given his shirt if he could have known exactly where he was just once a day!

Estimating Speed

Speed through the water is a deceptive matter to estimate. On yachts equipped with logs or speedometers which have been accurately calibrated, the problem is simplified. Beware, however, of relying exclusively on any piece of electronic or mechanical gear, for the ocean environment has a way of corroding and making inoperable even the best of equipment. It is a good idea to develop a sense of speed.

A man can walk comfortably at about three miles per hour. Using this fact as a basic reference one can with practice learn to judge the speed of a boat through the water quite accurately.

It is most important that one gain proficiency in judging boat speed under various conditions. Often sailors overestimate their speed on the wind and underestimate it when running free. This is probably because the heel of the boat when close-hauled gives a feeling of speed, while sitting upright when running and with the apparent wind reduced makes one feel that the wind has lessened and that the boat has slowed down.

Another interesting method of measuring your boat's speed through the water is by noting with a stopwatch the seconds required for a chip of wood, crumpled paper towel, or the like to travel a given distance. The procedure here is to measure either 30 or 40 feet along the deck forward from the helmsman. One person then drops the marker over the side calling "mark" as it hits the water. The helmsman starts the stopwatch at that moment and stops it as the object passes him.

Obviously, the greater the distance which the marker travels the more accurate will be the results. The following tables give speeds in knots for both 30-foot and 40-foot distances.

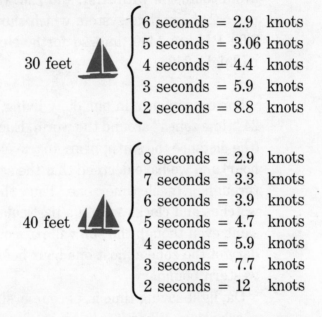

30 feet
$$\begin{cases} 6 \text{ seconds} = 2.9 \ \text{knots} \\ 5 \text{ seconds} = 3.06 \ \text{knots} \\ 4 \text{ seconds} = 4.4 \ \text{knots} \\ 3 \text{ seconds} = 5.9 \ \text{knots} \\ 2 \text{ seconds} = 8.8 \ \text{knots} \end{cases}$$

40 feet
$$\begin{cases} 8 \text{ seconds} = 2.9 \ \text{knots} \\ 7 \text{ seconds} = 3.3 \ \text{knots} \\ 6 \text{ seconds} = 3.9 \ \text{knots} \\ 5 \text{ seconds} = 4.7 \ \text{knots} \\ 4 \text{ seconds} = 5.9 \ \text{knots} \\ 3 \text{ seconds} = 7.7 \ \text{knots} \\ 2 \text{ seconds} = 12 \ \text{knots} \end{cases}$$

However one obtains his information it must periodically be plotted on the chart as a DR position, and it is the DR position which forms the basis for all celestial navigation.

Chapter Two: Time

Before going further, let us consider what time is and three different kinds of time which affect the navigator.

In essence, time is a measured portion of eternity. On earth we have chosen to use as our basic measuring interval the period of one revolution of the earth, or one day. The day, of course, we have divided into 24 equal parts, or hours. To avoid confusion with A.M. and P.M. time in navigation, we use a 24-hour time system which shows 1:00 P.M. as 1300, 2:00 P.M. as 1400, and so forth. Noon is thus 1200 and midnight 2400.

Zone Time

For convenience in our daily living, we have established 24 "time zones" around the world. Each of these is 15° wide (the distance the sun appears to travel across the sky in one hour) and we have decreed that the same time shall prevail throughout each time zone. Thus the time shown on all watches and clocks within a particular zone should be the same even though the sun will pass over the eastern boundary of the zone almost one hour before it passes over the western boundary.

Daylight-saving time is, of course, simply a modified form of zone time. When it is in effect, clocks are set forward one hour so that there will be longer periods of light in the evening. This results in a reduction by one hour of all zone descriptions. For example, the zone description for W 73° on standard time would be 5. On daylight-saving time, it would be 4.

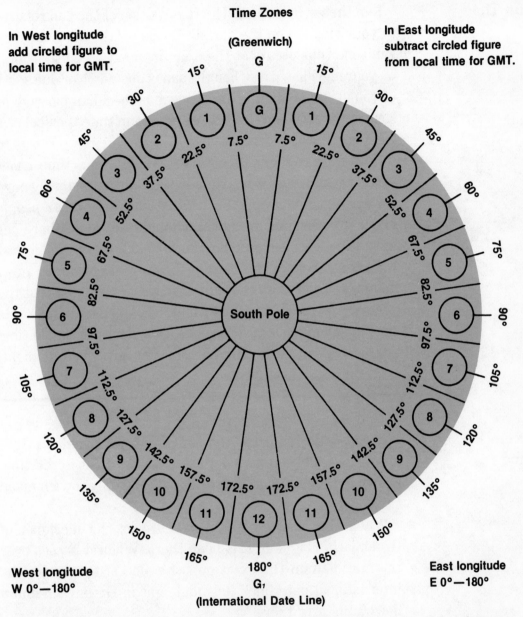

This is a diagram of the 24 time zones as shown on a cross section of the world as viewed from the South Pole. Each 15° meridian marks the center of a time zone. Each zone extends 7.5° on each side of its 15° meridian. A new day begins at Greenwich when the sun crosses G_1, the international date line. Hours of difference from Greenwich time are shown in circles. This difference is called the ZONE DESCRIPTION.

Figure 2.

Sun Time

Sun time is determined by the position of the sun relative to your location on the surface of the earth. Local noon by sun time thus occurs at the exact moment when the sun is on your meridian, i.e., when the sun is due north or due south of you. This moment is also known as meridian passage or local apparent noon (LAN). All other sun time is calculated from that basic reference.

If we were to regulate our lives strictly by sun time, chaos would result for, as we have seen, the sun is constantly moving across the heavens. For this reason, only other people in our precise same meridian of longitude would share our exact time. This is impractical because those a few miles east would see the sun rise, and meridian passage occur, a few minutes earlier than those further west (assuming that the horizon were unobstructed by hills, trees, and buildings). Sun time in any given location is thus a very precise sort of time as the sun can be in only one place at any given moment.

Greenwich Time

Greenwich time is the time in Greenwich, England. It is a form of timekeeping most useful to us, for it is the time shown in the *Nautical Almanac* as GMT—Greenwich mean time.

When the "average" sun passes over the meridian of Greenwich, England, it is noon there. When it passes over the meridian on the exact opposite side of the world (the international date line), it is midnight in Greenwich and a new day starts there.

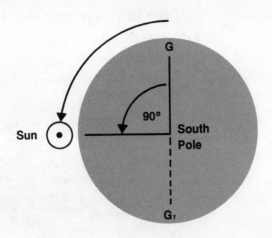

Figure 3.

In the diagram above, we are looking at the South Pole of the earth from a vantage point in space. The sun will always be depicted as traveling counterclockwise. If it is over G (Greenwich) at noon and will be over G_1 (international date line) at midnight, the diagram must represent 6:00 P.M. in Greenwich, or halfway between noon and midnight.

In the above diagram, it can also be seen that the sun has traveled one quarter of the way around the earth, or 90° of the circle.

If the sun were over G_1, the sun would be 180° from Greenwich.

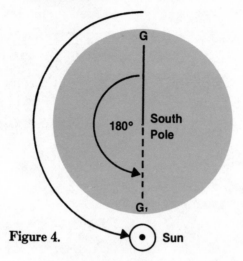

Figure 4.

Thus it may be seen that Greenwich time may also be expressed in terms of the degrees of a circle from the Greenwich meridian.

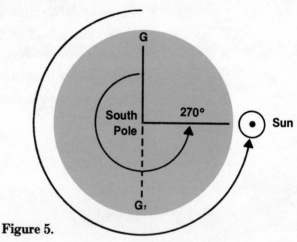

Figure 5.

This is the Greenwich hour angle (GHA) which the sun has traveled. Greenwich time may be converted into an equivalent number of degrees of angle, or arc, quite simply by using the first table in the back of the *Nautical Almanac*, "Conversion of Arc to Time." This table is reproduced in the back of this book. (See Appendix 1.)

Radio Time Signals

Radio time signals are broadcast day and night by the United States on 2.5, 5, 10, 15, 20 and 25 MHz with voice broadcasts every minute. Canada also broadcasts a strong signal on 6.3 megacycles and every five minutes on 3.330, 7.355 and 14.640 kHz. With these signals the navigator's watch can be set precisely to the correct time or any variation noted.

Taking the Time

As noted above, the sun is constantly moving across the sky at approximately 15° per hour. Since each hour at the equator represents approximately 1,044 miles, it is obvious that accurate time must be observed if one is to obtain a meaningful fix from a celestial observation. An error of *four seconds* in time will result in an error of *one mile* in location. Therefore, great care must be taken both in keeping and noting the time.

If someone is available to assist the navigator while he is shooting, it is helpful. The navigator can then prepare the timekeeper by stating, "Stand by"; and as he makes his final adjustment, add, "Mark." The timekeeper should note the second, then the minutes, and lastly the hour of the shot. He should then list beside the time the altitude reading from the sextant given him by the navigator.

Some navigators use a stopwatch as an aid in taking the time. The method here is to start the stopwatch at a convenient whole minute (which is noted on a work sheet) and punch it to stop and thus record the moment when the navigator calls,"mark." The time shown on the stopwatch is then added to the original whole minute time and a very precise result obtained.

Good results can be obtained without an assistant using the same methods but care must be taken so that the result will be as accurate as possible.

As one follows the sun for a noon sight, it will rise, appear to hang for a few moments at its zenith, and then begin to descend. Since it may be difficult to determine the actual second of its greatest height because of its apparent hesitation, the following will prove useful. During its ascendency, take and record as many shots as possible. Be sure to note both the time and the HS of each shot. Then as the sun begins its descent, set the sextant at the same altitude that it was on for one of the ascending shots, and record the time when the sun touches the horizon at this setting. Half time between the two shots will then be the moment of meridian passage.

Time of Meridian Passage

It is possible to follow the sun with a sextant beginning about 1130 local time and noting the zenith by taking almost constant shots up until the sun has clearly begun to descend. However, there is a simple method of predicting the time of meridian passage and its use will make long periods of squinting through the sextant's eyepiece unnecessary.

To predict the time of local apparent noon (LAN—also called meridian passage), begin with DR longitude and enter the GHA table for the date seeking the GHA nearest the value of your longitude but less than it. For example, on January 1, 1982, we are in the North Atlantic at DR longitude W 48°37'. Looking down the GHA column we find that at 1500 hours the sun's GHA is 44°06'. We then subtract this GHA from our longitude (first rounding off the fraction to its nearest whole number for simplicity), thus 48°37' − 44°06' = 4°31'.

Refer now to the table of "Increments and Corrections" and look down the Sun/Planet column until you find 4°27'. Reading to the left, the value of 4°31' is shown to be 18ᵐ04ˢ. This increment is then added to the 1500 hours for a total Greenwich time of 15ʰ18ᵐ04ˢ.

Reference to Figure 2 will show that at longitude W 48°, we are in the third time zone. Therefore, to convert from GMT to local time we must subtract 3 hours. Thus, $15^h18^m04^s$ – 3 hours = $12^h18^m04^s$. This should be the local time of meridian passage.

Since the DR longitude may be slightly off from the true longitude, it would be well to begin observations approximately 10 minutes before the predicted time of meridian passage and to continue them until the sun clearly has begun to ascend.

Maintaining a Time Log

Ten days or so before setting forth on an ocean passage, it is a good idea to prepare a simple log which will reflect any variation of the ship's best timepiece from an official time source. In many cities there is a telephone number which you may call and receive a recorded time signal accurate to Naval Observatory Time. Many commercial radio stations also broadcast accurate time signals, particularly at noon. The time signal should be obtained and noted at approximately the same hour each day, say lunch time or first thing in the morning.

The procedure here is to make two columns on a page with the date on one side and any watch error, as noted from the official time signal, beside it. If this is done for ten days prior to sailing, a pattern will emerge in which it will be apparent that the ship's watch or clock is gaining or losing a certain number of seconds per day. For instance, if you have kept such a log and it appears that your watch is gaining a second per day, you can extend your log for as long as you wish into the future by applying this same rate of change to future dates.

In this way in working out a sight you will need only to refer to the proper date in your time log and your watch error will be available at a glance.

Naturally, if you can obtain radio time signals while at sea, you should do so as these will eliminate any chance of even a small error becoming compounded into a large one with the passage of the days. See page 13 for the frequencies of the United States and Canadian time broadcasts.

Below is an example of a simple time log:

Date	Watch Error
6/1/82	00″
6/2/82	+01″
6/3/82	+02″
6/4/82	+03″
6/5/82	+04″
6/6/82	+05″
6/7/82	+06″
6/8/82	+07″
6/9/82	+08″
6/10/82	+09″
═══════════════════ Departure	
6/11/82	+10″
6/12/82	+11″
6/13/82	+12″
6/14/82	+13″
etc.	

Obviously if your watch error is +12 seconds, as reflected on your time log or obtained from a radio signal, you will reverse the + to a − on the second line of Step 2 of the work form to correct for the watch error.

Chapter Three: The Sextant

Basically, the sextant is merely an instrument with which one can measure angles with great accuracy. With it one can measure the angle between the horizon, himself and a celestial object. In essence, it is little more than a sophisticated protractor.

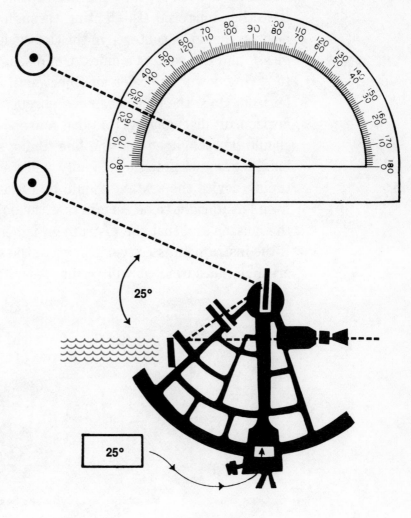

Figure 6.

If we could hold a protractor level and measure the angle of the sun with it, we would obtain the same information which the sextant gives us. The sextant, of course, enables us to measure much more accurately than we ever could merely with a protractor. The sextant is also a delicate instrument and must be handled with great care lest it be damaged.

Mounted in front of the telescope or sighting tube of a sextant is a piece of glass, one side of which is transparent, the other mirrored. In sighting through the sextant the horizon should be centered in the clear side of this "horizon glass" and the sextant's index arm disengaged and moved to "bring down" the celestial object until it just kisses the horizon. Once the object has been brought down near the horizon by disengaging the arm and moving it, the arm should be reengaged and the fine tuning adjustment used for the greatest precision. Finally, while adjusting the fine tuning device the sextant should be rocked in a slight arc from left to right to be certain that the object just touches the horizon and that the instrument is being held vertical. If the instrument is not vertical when the sight is taken, an error in measurement will result.

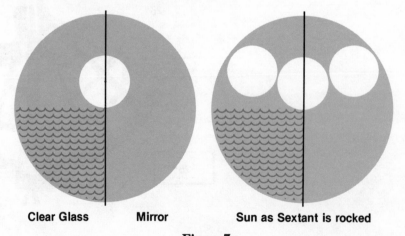

Clear Glass   Mirror   Sun as Sextant is rocked

Figure 7.

Normally, the bottommost edge of the sun or moon is brought to the horizon. This is known as the *lower limb* of the body. If we wish, we can use the topmost edge or *upper limb*. Stars and planets have no visible diameter and are thus merely brought to the horizon as points of light. See Figure 10.

The raw sight which is obtained directly from the sextant is commonly called the height shot (HS). The HS must be corrected for several factors before is it useful to us.

Instrument Error

In looking through the telescope or sighting tube of a sextant, the horizon may appear broken in the middle of the field of view when the index arm is set at zero. By careful adjustment the horizon can be made to become a single continuous line. When the horizon appears as a straight line, the sextant should read exactly zero. If it does not, note the amount of the error and whether it is above or below zero. If it is above zero, the instrument will be reading that much too high in any observation, and a correction must be made in the HS to account for this. This is known as instrument error (IE).

Obviously, if the sextant is giving readings that are too high, the error will be deducted from the reading of the instrument. If it is giving readings which are too low, the instrument error should be added.

Dip

Another necessary correction is for the dip or height of eye above the water. The horizon appears to dip away as one gets higher above the water so we must compensate for this.

There is a table of the correction for dip inside the front cover of the *Nautical Almanac*, and also as Appendix 4 to this book.

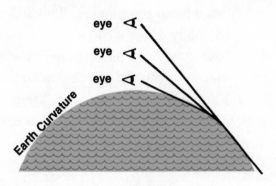

Figure 8.

Height of the eye is given in both meters and feet, therefore, be sure to use the correct scale. The height of your own eye should be added to the height of the deck above the water to determine the correct "height of eye" with which to enter the table. On most medium-sized yachts this will total approximately 10 feet, for a dip correction of 3.1 minutes of arc. The dip correction is always subtracted from the HS. Part of this table is shown below.

ION TABLES 10°-90°—SUN, STARS, PLANETS

Upper Limb	STARS AND PLANETS				DIP			
	App. Alt.	Corrⁿ	App. Alt.	Additional Corrⁿ	Ht. of Eye	Corrⁿ	Ht. of Eye	Ht. of Eye	Corrⁿ
′	° ′	′	**1972**		m		ft.	m	′
21·2	9 56	−5·3	**VENUS**		2·4	−2·8	8·0	1·0 − 1·8
21·1	10 08	−5·2	Jan. 1–Feb. 29		2·6	−2·9	8·6	1·5 − 2·2
21·0	10 20	−5·1	°	′	2·8	−3·0	9·2	2·0 − 2·5
20·9	10 33	−5·0			3·0	3·1	9·8	2·5 − 2·8
20·8	10 46	−4·9	42	· 0·1	3·2	−3·2	10·5	3·0 − 3·0
20·7	11 00	−4·8	Mar. 1–Apr. 15		3·4	−3·3	11·2	See table
20·6	11 14	−4·7	0°	+0·2	3·6	−3·4	11·9	←

Refraction

Refraction is another correction that must always be applied to one's HS. Light rays entering the earth's atmosphere from a celestial body bend as they approach earth just as a stick appears to bend when thrust into the water.

The lower the angle at which light rays enter the atmosphere the greater the bending.

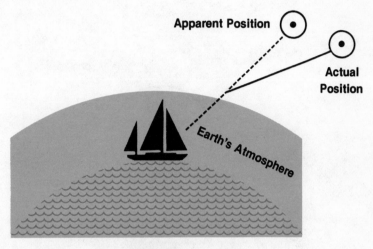

Figure 9.

The first tables inside the front cover of the *Nautical Almanac*, "Altitude Correction Tables 10°–90°" and "0°–10°" give the corrections for refraction. The portion of this table for altitudes 10°–90° is reproduced herein as Appendix 4. Be sure to use the proper column, Oct.–Mar. or Apr.–Sept., and enter the table with the value of your apparent altitude (App. Alt.), that is, your HS corrected for dip and instrument error. This table also makes allowance for the fact that the center of the sun is not sighted but either its lower or upper limb is brought to the horizon. Therefore, care must also be used to make sure that the proper column for upper or lower limb is used. Lower limb sights always give a + correction; upper limb sights always give a – correction.

The upper limb sight is most often used with the moon when it is in a crescent phase, but it will occasionally be useful with the sun in the event its lower limb is obscured by a cloud.

These three corrections must always be made to any sextant readings. When they have been applied, the HS becomes the height observed (HO).

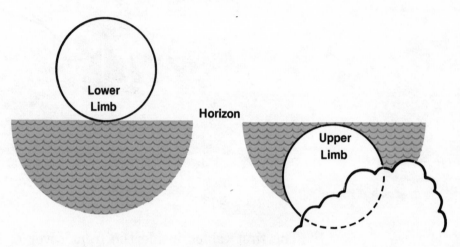

Figure 10.

The following form will be useful in finding the total correction to apply to the HS to obtain HO:

HS   _____
IE (+ or –)   _____
Dip (–)   _____
_____

App. Alt.   _____
Alt. Corr. (+ or –)   _____
_____

HO   _____

Example: Find HO from HS (sextant reading) of 43°16′ on lower limb, height of eye 10 feet with instrument error of +3′, date January 1, 1982.

HS	43°16′
IE	–03′
Dip	–03.1

Enter Altitude Correction table with this argument

App. Alt.	43°10′
Alt. Corr.	+ 15.2

HO	43°25′

Artificial Horizon

If the open sea is not readily available for practice sights, a shallow pan of water may be used as an artificial horizon. Place the pan so that the sun's image can be seen reflected in it. Sight the image in the water in the clear side of the frame and bring down the sun in the sky in the usual way. Using this system, however, the two suns are superimposed on each other to form one disc instead of one being brought to the horizon. No correction for dip is used; other corrections as usual. HS is then halved to obtain apparent altitude and from this HO is obtained. If the wind disturbs the water too much, light oil, such as kitchen cooking oil, may be substituted.

Chapter Four:
Latitude and Longitude

Latitude

If you will examine a world globe, you will notice the equator as the large band circling the earth midway between the North and South poles. Parallel to the equator and also circling the earth above and below it are additional bands representing the degrees of latitude. These increase in value as one moves either north or south from the equator in a range from 0°, the equator, to 90° at either the North or South poles. Each degree of latitude is made up of 60 minutes and each minute of 60 seconds. It is thus possible to state precisely how far north or south of the equator one is located in terms of degrees (°), minutes (′) and seconds (″) north or south.

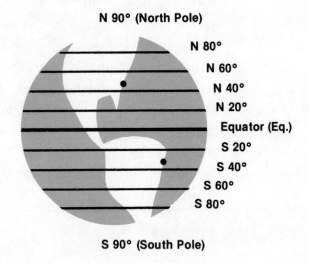

Figure 11.

For example, Washington, D.C., is located at N 38°50′, New York lies at N 40°30′, and Buenos Aires, Argentina at S 34°40′.

Latitude from the Noon Sight

Declination (dec.) is simply the number of degrees north or south of the equator at which a celestial body may be found. When it is summer in the Northern Hemisphere, the sun has a north declination and it is winter in the Southern Hemisphere.

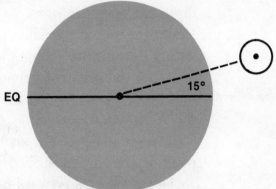

Here the Sun has a declination of 15° North

Figure 12.

As we have seen, the *Nautical Almanac* gives the declination of the sun and moon for each whole hour of GMT throughout the year. Since declination changes much more

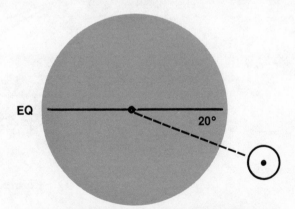

Here the Sun has a declination of 20° South, thus it is summer in the Southern Hemisphere and winter in the Northern Hemisphere.

Figure 13.

slowly than the hour angle, it is not always necessary to interpolate between the hours for minutes and seconds, but it is sufficient simply to use the declination given for the nearest whole hour to the time of your sight.

At local apparent noon (LAN) we can obtain latitude directly from a sun sight. To do this we need know only the HO of the sun at meridian passage and its declination. If, with a sextant, we follow the sun as it ascends in the sky and note the greatest value that it achieves before beginning to descend, we can determine our latitude with only the *Nautical Almanac*.

At the moment of our shot we are actually standing at a point on the earth's surface (L — latitude), and the horizon we see as we look forward is a line which is tangent to the earth circle at that point.

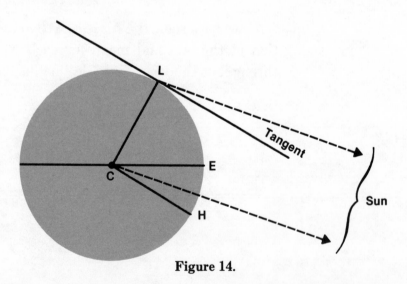

Figure 14.

The sun is so far from earth that there is no appreciable difference in an observation of the sun, a star or planet from either L or C, the center of the earth. Therefore, we can draw C H parallel to the tangent line and use C H in our diagrams rather than the tangent line, as this makes the diagram simpler. Since two parallel lines crossing a straight line form equal angles, the use of C H rather than the tangent line will not affect the rest of the diagram.

For example, let us suppose that it is January 1, 1982, and that we are somewhere off the East Coast of the United States. Let us assume that the highest corrected reading we obtained from our sextant, our HO, as we followed the sun up to meridian passage was 18°26'. We can now construct a simple diagram which will graphically demonstrate how to obtain our latitude.

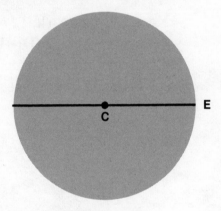

Figure 15.

We will first draw a circle with a line running through it representing the earth and the equator, with a dot representing the center of the earth. Let us assume that all angle lines to be drawn will terminate at the earth's center, C.

Next, we add the sun, S, in the location of its approximate declination either north or south. This declination we obtain from the daily pages of the *Nautical Almanac*. For January 1, 1982, at 1200 the declination is shown as S 23°00.3'. We shall drop the fraction of a minute for convenience and sketch in the sun at an angle below the equator (since the declination is south) of approximately 23°00'.

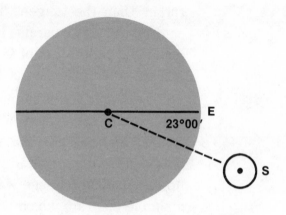

Figure 16.

Next, place a mark, L, on the diagram at approximately your dead reckoning (DR) latitude. If you bear in mind that the topmost and bottommost points of the circle represent the North and South poles and that there are 90° between

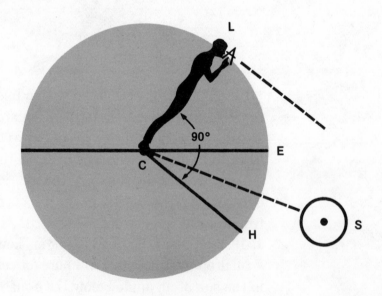

Figure 17.

each of them and E, you should have no problem marking this approximate latitude. For example, 45° is halfway between a pole and the equator, 30° is one-third of the distance, and so forth.

Now draw a line from L to C and then a line at a 90° angle to this line from C to the earth circle. Mark the point where this last line meets the earth circle H. This line represents your horizon.

The final step in the construction of the diagram is to place in the angle SCH the HO you obtained from your sextant reading.

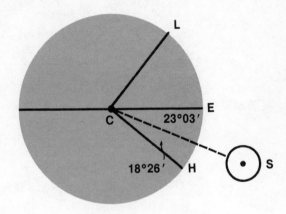

Figure 18.

Simple addition and subtraction will now give us the latitude. We know the angle LCH contains 90° (we drew it this way). We are seeking the angle LCE as the degrees of this angle actually represent our latitude. We need then to add together the two known angles,

ECS = 23°03′
SCH = <u>18°26′</u>
ECH = 41°29′

N Lat S Dec

90 − (HO+ Dec) = Lat

29

and subtract them from 90° to obtain the angle LCE. We can conveniently express 90° as 89°60'. Thus

$$89°60'$$
$$- \underline{41°29'}$$
$$48°31' \quad = LCE - \text{This is our latitude.}$$

If we were north of the equator and the declination of the sun was North at 10°, our diagram would look like this with a sextant reading of 50°.

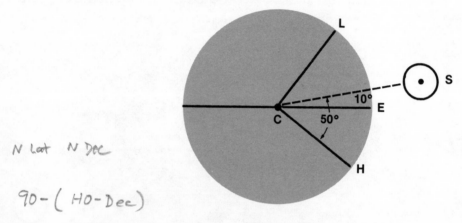

N Lat   N Dec

90 – (HO – Dec)

Figure 19.

Here again our objective is to find LCE, as this is our latitude. We subtract the value of SCH from 90° and obtain the value of LCS.

$$90°$$
$$- \underline{50° \text{ SCH}}$$
$$40° \text{ LCS}$$

We then add SCE to this to obtain LCE.

$$40° \text{ LCS}$$
$$+10° \text{ SCE}$$
$$50° \text{ LCE — This is our latitude.}$$

If we were south of the equator the only change is to place H beyond the sun as we face it.

In essence, all we must do in each case is obtain the value of the angle LCE. The diagram may be turned in several different ways but always LCH will be a 90° angle, and we will have two known angles which will form portions of it. By combining the known angles of the declination and sextant reading (HO) to suit our particular circumstances, we will always be able to find the value of LCE, which will give us our latitude.

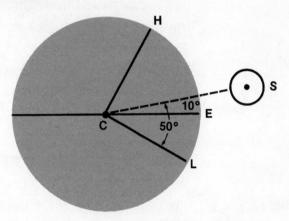

Figure 20. S Lat N Dec

90 − (HO − Dec)

Remember that if we are facing south at the time of the sight, H will be south of us, below the equator as in Figures 4 and 25. If we are facing north, H will be north of us as in Figure 26 above. Thus, L and H will always straddle the sun line.

To avoid confusion it is important always to draw a diagram so that the relationship of the various angles will be clear. Also, this will eliminate the need for memorizing a series of rules which tell what to do when north or south of the equator when declination is north or south. The diagram shows all simply and clearly, so always draw it and be sure.

In making your diagram it is not necessary to draw it with great precision, measuring angles with a protractor and such. A freehand drawing is sufficient since the diagram is merely a tool to help you visualize the relationship of the various angles and bodies. It is useful, however, to make the diagram large enough so that the numbers of the angles can be written in the actual angles and be easily read. When you are wet, cold, tired, and hungry, every device that will lessen possible confusion and the resultant possible errors is important. For ease of use and clarity, an earth circle of at least three inches in diameter is recommended.

Plotting

In plotting on a chart the latitude obtained by your noon sight calculations, measure with your dividers either north or south from the nearest whole degree parallel crossing the chart the required minutes obtained from the scale on the side of the chart. Do the same to plot the longitude except that with longitude you must measure east or west from a meridian the minutes obtained from the scale at the top or bottom of the chart.

In measuring nautical miles on a chart, always be sure to use the scale at the side rather than at the top or bottom. The scales on the side show latitude, the lines of which are parallel, and they are thus uniformly spaced. The top and bottom scales show longitude, the meridians of which are not parallel, and which therefore cannot be used in measuring miles.

Problems

Determine latitude as follows:

1. January 2, 1982 DR Lat. N 42°50'
   Time of LAN 12:10:05
   Sextant reading HO 24°59'

2. January 3, 1982 DR Lat. S 30°02'
   Time of LAN 11:59:22
   Sextant reading HO 83°14'

(Answers on page 104.)

Longitude

Distance around the world, east and west, is called longitude. It is measured east or west of Greenwich, England, which marks the zero meridian. Meridians are imaginary lines which radiate from the North Pole like spokes in a wheel and extend due south to the South Pole.

Values of meridians increase as one goes east or west from Greenwich until one reaches the international date line, which is 180° from Greenwich. Each degree (°) of longitude is composed of 60 minutes (') and each minute of 60 seconds ("). It is thus possible to state with precision exactly how far east or west one is from the Greenwich meridian in terms of longitude east or west.

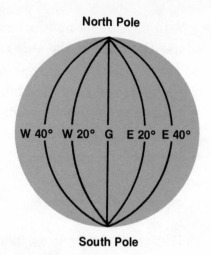

North Pole

W 40° W 20° G E 20° E 40°

South Pole

Meridians of Longitude Side View of Earth

Figure 21.

The angle between Greenwich, the center of the earth, and the sun is constantly changing as the sun circles the earth. This angle is known as the Greenwich hour angle (GHA) and is stated in continuously increasing values from zero at Greenwich to 359°+ as the sun approaches Greenwich from the east. Longitude, on the other hand, is never expressed in values of over 180° as it switches from west to east at the international date line and at Greenwich. In all the following diagrams we are viewing the South Pole of the earth, and the sun is traveling counterclockwise.

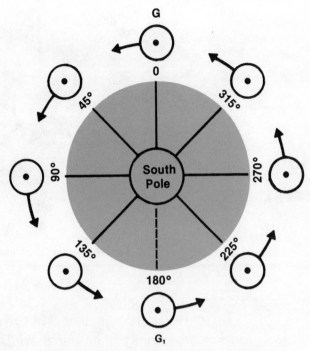

Various Greenwich Hour Angles of the Sun

Figure 22.

The sun travels at approximately 15° each hour so that in each 24-hour day it makes one complete circle of 360° around the earth.

If the sun traveled at a uniform rate of speed, it would cross the Greenwich meridian at precisely 12:00 noon each day. Unfortunately, the sun does not travel at a uniform speed, and so from day to day there are variations in the exact time it crosses the Greenwich meridian. This variation from the average is called the equation of time, and it represents the difference between the average time of meridian passage and the precise time when it occurs on a particular day.

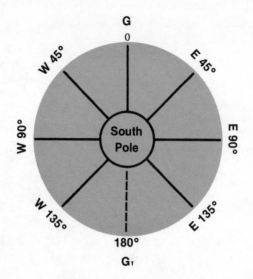

Longitudes, East and West

Figure 23.

By meridian passage is meant the exact moment when the sun or other body is on our meridian; i.e., when it lies due south or due north of us. Up until that moment the body will have been ascending in the sky. After it, it will be descending.

The sun is at its highest point in the sky and will be either due south or due north of the observer as he faces it at local apparent noon.

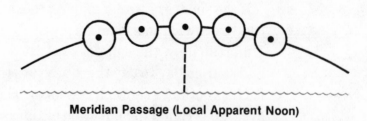

Meridian Passage (Local Apparent Noon)

Figure 24.

If you will examine a typical page from the *Nautical Almanac*, as reproduced in Appendix 3, you will note the first column on the right-hand page to be GMT, Greenwich mean time.

The second broad column is headed Sun and is divided into two parts, GHA and Dec. The term Dec. refers to declination and is explained in the section on latitude. GHA means Greenwich hour angle. You will note that the GHA is always at its lowest near 1200. This is because the sun would be directly over Greenwich at noon but for the equation of time.

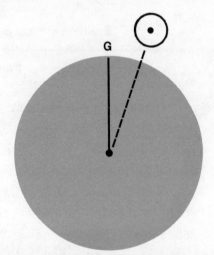

Equation of Time places Sun late at 1200

GHA = 358°41′

Figure 25.

If we had an accurate timepiece on board our vessel and were able to time the precise moment of meridian passage of the sun, we could by a simple computation establish our longitude.

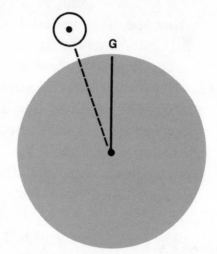

Equation of Time places Sun early at 1200
GHA = 3°57′

Figure 26.

For example, on January 1, 1982, we know that we are somewhere in the North Atlantic and our DR gives us a longitude of 63° West. With the sextant we take the time of the sun's meridian passage at 12ʰ03ᵐ21ˢ.

It is now important to determine what time it is in Greenwich. As shown in Figure 2, the world is divided into 24 time zones. Each time zone is 15° wide and extends 7.5° on either side of a 15° meridian.

Since our DR places us at approximately 63° West, we can see from Figure 2 that we are in the 4th zone, or that it is exactly four hours later in Greenwich than it is where we are. Thus the GMT of our meridian passage was 12ʰ03ᵐ21ˢ + 4 hours or 16ʰ03ᵐ21ˢ.

Now look down the GMT column in the *Nautical Almanac* for January 1, 1982 (Appendix 3A), to the 16th hour and note the GHA. It is shown as 59°05.7′. Since the time shown is for the whole hours, we must add to the value shown for the 16th hour the value of the additional 03ᵐ21ˢ.

The second table in the back of the *Almanac* is entitled "Increments and Corrections." Selected pages from this table are reproduced herein as Appendix 2. Each page gives the values for 2 minutes of time in separate columns and shows also the cumulative value of each additional second. To use this table simply turn to the page with the number of minutes with which we are concerned (in this case 3) and run down the Sun/Planet column to 21 seconds. This is shown as 0°50.3'. This increment is then added to the value obtained for the even hour, and we have our longitude.

59°05.7' [Whole hour GHA]
+ 0°50.3' [Increment for 3m21s]
————————
59°56'   West — this is our longitude

If you were in east longitude at the time of your sight it would be necessary to convert your GHA into east longitude by subtracting your GHA from 360°. Remember that GHA is expressed as an angle increasing continuously from zero at Greenwich to 359°+ as Greenwich is approached from the east. Since longitude is never expressed in terms of over 180° it is necessary to subtract any GHA of more than 180° from 360° to obtain east longitude. Thus, if the sun's GHA were 270° at the time of your sight, you would be at 90° East. See Figures 22 and 23.

Problems

Determine longitude from the following:
1. January 2, 1982 DR Long. W 121°16'
   Time of LAN 12:10:05
2. January 1, 1982 DR Long E 61°08'
   Time of LAN 11:59:22
(Answers on page 105.)

Chapter Five: Moon Shots

From time to time during daylight hours both the moon and the horizon are visible. At such times, a fix may be obtained from the moon's meridian passage by following the same procedures as for the sun with but one additional correction to the HS. Also, of course, GHA and Dec. are used as given for the moon in the *Nautical Almanac*.

Because the moon and the earth are so close together, astronomically speaking, it does make a difference that observations cannot be made from the earth's center. This difference of angle of observation from the earth's surface rather than its center is known as parallax.

The last column in the moon section of the daily pages of the *Almanac* is headed HP. This stands for horizontal paralax and is an additional correction factor which must be applied to HS to obtain HO.

In working a moon sight note the time of meridian passage and the HS as usual. In addition, note the HP for the nearest whole hour to the time of the sight. Turn then to the table "Altitude Correction for the Moon" inside the back cover of the *Nautical Almanac*. The correction is in two parts. The first is similar to that for the sun as found in the front of the *Almanac* and is entered with apparent altitude. In the second, the lower part of the table is used with the given HP applied in the same column vertically as that from which the first correction was taken.

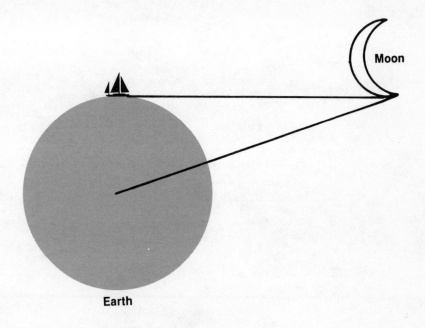

Figure 27.

The two corrections from the upper and lower portions of the table are then added together and the total correction added to the apparent altitude. This then becomes the HO.

In the event that the upper limb has been used instead of the lower limb, 30' is to be subtracted from the altitude of the upper limb.

Further instructions for the moon sight appear with the tables in the back of the *Nautical Almanac*.

PART TWO

SUN LINES ANY TIME

Introduction

As useful as the noon sight is, it has a major limitation. If the navigator should be occupied with some other matter at the moment of meridian passage, or if either the sun or the horizon should be obscured by a cloud or haze just then, the opportunity for the noon sight will be lost until the following day. Also, if one were closing with a coast an accurate fix as late in the afternoon as possible would be desirable over waiting until noon the following day.

For these reasons, it is well to be able to go beyond the noon sight and obtain a fix, or at the very least a line of position, at any time when the sun and the horizon make a proper observation possible.

The Navigator's Notebook

Many navigators like to compile a three-ring notebook in which they can save choice articles on navigation, work forms, plotting sheets, a time log, and even the ship's log. This is a good idea and is recommended to you.

It is a good idea also to take and time accurately several sun shots each morning and several more each afternoon. Generally, it is not necessary to work all of these out or plot them all. However, it will be comforting to the navigator to have this information available just in case a later shot puts him a thousand miles up the Amazon River when he hoped he might soon be approaching the big island of Hawaii! This will be especially so if clouds pick this time period in which to hide the sun for a three-day rest. Having additional observations to fall back on will then be most helpful.

Neatness and accuracy should be the navigator's watchwords. In addition, of course, all items should be labeled so that a problem will speak for itself and there will be no need to rely on one's memory, which may become unclear with the passage of time or the arrival of additional information.

Chapter Six:
Determining Location

Line of
Position
(LOP)

Any time an accurate observation of the sun's altitude can be made and timed, it is possible to obtain a line of position (LOP). One LOP will not yield a complete fix, but it will tell the navigator a great deal. A single LOP tells the navigator that he is somewhere on that line. If he can establish a second LOP which intersects the first at a reasonably large angle, it will tell him that he is somewhere on that second line. If the two LOPs cross each other in the general vicinity of his DR position, it is reasonably safe to assume that a proper fix has been obtained.

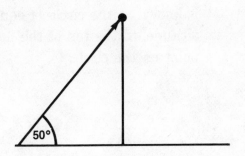

Figure 28.

Imagine that there is a flagpole in the middle of a level field and that you can measure the angle from the ground at your feet to the top of the pole. You determine it to be 50°.

Now assume that you measure the distance between you and the flagpole and find that you are exactly 26 feet from it. If you were to stand 26 feet from the pole in any direction and measure the angle to its top, it would always be the same, in this case 50°.

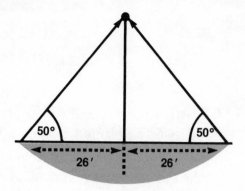

Figure 29.

Viewed from above, you could be standing anywhere on a circle with a radius of 26 feet, as in Figure 30. This circle is known as the circle of equal altitude since the angle, or altitude, to the top of the pole will be the same from any point on the circle.

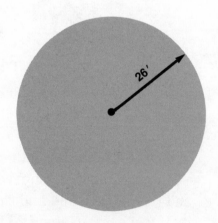

Figure 30.

If you were to move farther away from the flagpole, the angle to its top from your new position would become smaller.

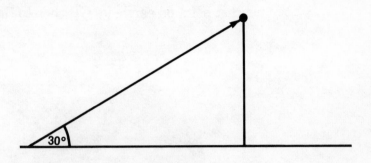

Figure 31.

If you were to move closer to the flagpole the angle would increase until it reached 90° as you stood at its base and looked straight up.

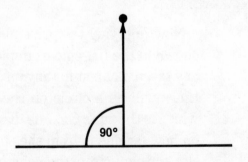

Figure 32.

By using information from the *Nautical Almanac* together with tables contained in *Sight Reduction Tables for Air Navigation, (Pub. No. 249)*, we can use the above principle to determine a precise LOP at sea by measuring with a sextant the angle between the horizon, our ship, and the sun.

As the sun circles the earth it is always directly above some particular spot on the planet. This spot is known as its geographical position (GP). This is demonstrated in Figure 33 by the broken line which descends from the sun to the nearest point on earth, its GP, then to the center of the earth.

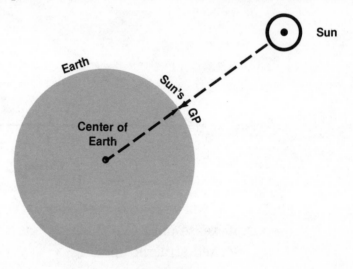

Figure 33.

Now, applying the principle of a circle of equal altitude as shown in the flagpole examples above, it is apparent that at any given moment for any measurement of the sun's altitude, there will be a circle on the earth's surface, the center of which is the sun's GP. The diameter of this circle will depend on how far one is from the sun's GP. If the sun is low in the sky, the angle from the point of observation to it will be quite small, and the circle of equal altitude will thus be very large. If the sun is high in the sky, the angle to it will be quite large and the circle of equal altitude will be much smaller. This is shown in Figure 34.

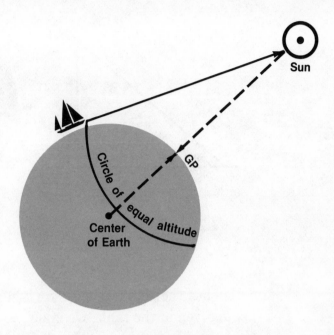

Figure 34.

If you were to go into a field with a sextant and a compass and carefully measure the angle to the top of a flagpole as 50° and the compass bearing from you to the pole as 180°, you could put a dime on the ground and return to find it an hour, a week or a month later. You could be at only one spot.

So it is with the sun and the tables. What the *Nautical Almanac* together with Pub. No. 249 basically say to us is,

"Assume that you are at a specific position at a specific time. If you are in fact at that position at that time, the angle you measure between the horizon and the sun should be X° and the true bearing towards the sun's GP will be Y°."

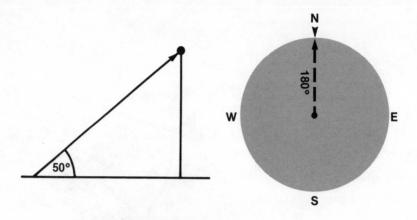

Figure 35.

If in fact the angle we measure is slightly greater than the angle given in the table, we know we are slightly nearer to the sun's GP than our assumed position. If the angle is less, we are farther away. The tables will also yield the true (*not* magnetic) bearing from our assumed position to the sun's GP. This bearing is known as azimuth and is dealt with in more detail later.

Simply stated, this is all there is to establishing an LOP at sea using observations of the sun.

The nearer the observer gets to the sun's GP, the greater the observed angle will be and, consequently, the smaller the circle of equal altitude will become, as shown in Figure 36. Just like the flagpole!

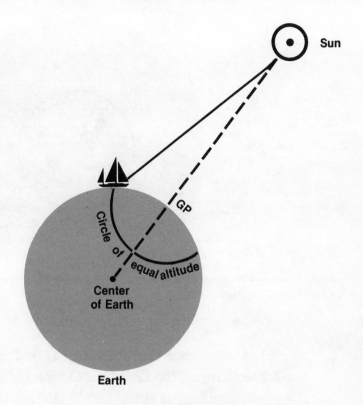

Figure 36.

As in the examples with the flagpole, an observation of the sun made from any location on this circle at the same instant will yield the same altitude angle.

Viewed from somewhere in space directly above the sun's GP the diagram would look very much like our familiar flagpole, as viewed from above.

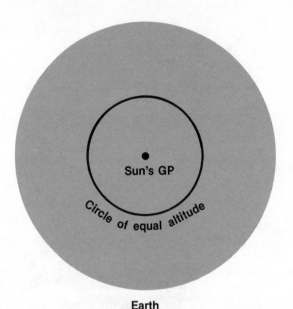

Sun's GP

Circle of equal altitude

Earth

Figure 37.

Let us assume that it is July 4, 1982, and we are sailing somewhere off the East Coast of the United States. Our DR position places us at latitude 37°15′ North, longitude 75°22′ West. At exactly $08^h32^m25^s$ on the ship's chronometer we take a shot of the sun with the sextant and bring its lower limb into the barest contact with the horizon, as is described in Figure 10, Part 1 above. We estimate our height of eye at 10 feet above the water. The uncorrected sextant reading, or HS, is 30°13′. We note an instrument error of +2′. Assume that there is no watch error.

Referring to Figure 2, we note that at 75° West longitude we have a 5-hour time difference from Greenwich. We also note that in west longitudes we *add* the zone description, or time difference, to local time to convert to GMT, which is the time shown in the *Almanac*. In our example then, we should add 5 hours to the time of our shot. Since it is summer, however, and daylight saving time is in effect, we reduce all

zone descriptions by 1 hour and therefore only add 4 hours.
Thus: $08^h32^m25^s$

$$\frac{+\ 4^h}{12^h32^m25^s}$$

This is the correct GMT of our sun shot.

The first step in solving our location problem is to open the 1982 *Nautical Almanac* to the daily page for July 3, 4, and 5 (reprinted in full as Appendix 3(B) hereof). We note that July 4 is a Sunday and confirm the date and day to be correct by reference to the ship's log. Always take the trouble to be sure you are reading the table for the proper day!

Refer now to the Sun column for July 4 at the 12th hour. We use the 12th hour because this is the nearest whole hour earlier than our shot and only whole hours are listed in the *Almanac*. For easy reference you may now wish to draw a line in the *Almanac* right under the 12th hour. While it is a good practice, and helpful, to underline pertinent hourly references in the daily pages of the *Almanac*, one should *never* underline any portion of the permanent tables as this may prove to be confusing in subsequent usage.

Note that the GHA whole hour value is given as 358°55.5′ and that the declination given for the sun is N 22°52.9′. As described on page 38, we must again add to the whole GHA of 1200 the value represented by the additional 32^m25^s. To do this, turn to the Increments and Corrections table in the back of the *Nautical Almanac* (Appendix 2 hereof) to the page for 32^m. Run down the Sun/Planets column to 25^s and note the value, which is 8°06.3′. This value is then added to that of the whole hour for the total GHA of our shot. Thus:

$$\frac{\begin{array}{r} 358°55.5' \\ +\ \ \ 8°06.3' \end{array}}{366°61.8'\ =\ \text{GHA}}$$

1982 JULY 3, 4, 5 (SAT., SUN., MON.)

G.M.T.	SUN G.H.A.	Dec.	MOON G.H.A.	v	Dec.	d	H.P.
3 00	178 59.6	N23 00.3	37 15.8	13.2	S17 50.0	7.4	54.2
01	193 59.5	23 00.1	51 48.0	13.1	17 57.4	7.3	54.2
02	208 59.4	22 59.9	66 20.1	13.1	18 04.7	7.2	54.2
03	223 59.3	59.7	80 52.2	13.1	18 11.9	7.1	54.2
04	238 59.2	59.5	95 24.3	13.1	18 19.0	7.0	54.2
05	253 59.0	59.3	109 56.4	13.0	18 26.0	7.0	54.2
06	268	N22 59.1	124		S18 33.0		
07		58.9		13.0	18 39		

	MOON G.H.A.	v	Dec.	d	H.P.
23	148 57.1	55.7	11 27.3 12.4	S20 23.4	5.3 54.0
	163 57.0		25 58.7 12.4	20 28.7	5.2 54.0
4 00	178 56.9 N22	55.5	40 30.1 12.3	20 33.9	5.1 54.0
01	193 56.8	55.3	55 01.4 12.3	20 39.0	5.0 54.0
02	208 56.6	55.1	69 32.7 12.3	20 44.0	4.9 54.0
03	223 56.5	54.9	84 04.0 12.2	20 48.9	4.8 54.0
04	238 56.4	54.7	98 35.2 12.2		4.7 54.0
05	253 56.3	54.4	113 06.4 12.2	S20 53.7	4.7 54.0
06	268 56.2 N22	54.2	127 37.6 12.1	20 58.4	4.5 54.0
07	283 56.1	54.0	142 08.7 12.1	21 03.1	4.4 54.0
08	298 56.0	53.8	156 39.8 12.1	21 07.6	4.4 54.0
S 09	313 55.9	53.6	171 10.9 12.0	21 12.0	4.2 54.0
U 10	328 55.7	53.4	185 41.9 12.1	21 16.4	4.2 54.0
N 11	343 55.6	53.2	200 13.0 12.0	S21 20.6	4.1 54.0
D 12	358 55.5 N22	52.9	214 44.0 11.9	21 24.7	4.1 54.0
A 13	13 55.4	52.7	229 14.9 12.0	21 28.8	3.9 54.0
Y 14	28 55.3	52.5	243 45.9 11.9	21 32.7	3.9 54.0
15	43 55.2	52.3	258 16.8 11.8	21 36.6	3.7 54.0
16	58 55.1	52.1	272 47.6 11.9	21 40.3	3.7 54.0
17	73 55.0	51.8	287 18.5 11.8	S21 44.0	3.5 54.0
18	88 54.9 N22	51.6	301 49.3 11.8	21 47.5	3.5 54.0
19	103 54.7	51.4	316 20.1 11.8	21 51.0	54.0
20	118 54.6	51.2	330	21 54.3	
21	133	51.0		21	
22		50.7			

	SUN G.H.A.	Dec.	MOON G.H.A.	v	Dec.	d	H.P.
17	43 52.6		32 01.1 11.4		1.3 54.0
18	58 52.5	N22 46.4	246 31.5 11.4	22 38.3	1.3 54.0
19	73 52.4	46.1	261 01.9 11.3	22 39.6	1.2 54.0
20	88 52.2 N22	45.9	275 32.2 11.4	S22 40.8	1.0 54.0
21	103 52.1	45.6	290 02.6 11.4	22 41.8	0.8 54.0
22	118 52.0	45.4	304 33.0 11.3	22 42.8	0.8 54.0
23	133 51.9	45.2	319 03.3 11.3	22 43.6	0.6 54.0
	148 51.8	44.9	333 33.6 11.3	22 44.4	0.6 54.0
	163 51.7		348 03.9 11.3	22 45.0	0.5 54.0
	S.D. 15.8	d 0.2	S.D. 14.7	14.7	14.7

Twilight / Sunrise / Moonrise

Lat.	Naut.	Civil	Sunrise	3	4	5	6
N 72				20 31	21 18	22 31	23 11
N 70			00 39	19 49	20 39	21 44	22 28
68			01 46	19 21	20 12	21 14	21 59
66			02 20	19 00	20 12	20 51	21 37
64			02 45	18 42	19 51	20 32	21
62		01 10	03 04	18 28	19 34	20 17	
60			03 21	18		20	
N 58			34				

Lat.	Naut.	Civil	Sunrise				
45	06 28	07 05	07 39	14 50	15 29	15 56	16 45
		07 21	07 59	14 36	15 13	15 35	16 24
S 50	06 39	07 28	08 08	14 19	14 53	15 25	16 13
52	06 44	07 36	08 19	14 11	14 44	15 13	16 02
54	06 50	07 44	08 31	14 02	14 33	15 00	15 49
56	06 56	07 54	08 45	13 52	14 22	14 45	15 34
58	07 09	08 05	09 02	13 40	14 08	14 27	15 16
S 60				13 27	13 52		

Twilight / Moonset

Lat.	Sunset	Civil	Naut.	3	4	5	6
N 72				23 46	00 28	00 38	01 07
N 70				00 25	00 57	01 19	01 54
68				00 45	01 19	01 45	02 25
66	23 24			01 02	01 36	02 06	02 48
64	22 21			01 15	01 51	02 23	03 06
62	21 47	22 56		01 27	02 04	02 38	03 22
60	21 24	22 15		01 38	02 04	02 51	03 3
N 58	21 03	21 48	23 02	01 47	02 15	03 02	
56	20 47	21 27	22 24	01 5	25	03 12	
54	20 34		21 59				
52	20 22						

Day	SUN Eqn. of Time 00h	12h	Mer. Pass.	MOON Mer. Pass. Upper	Lower	Age	Phase
	m s	m s	h m	h m	h m	d	
3	04 01	04 07	12 04	22 13	09 49	12	
4	04 12	04 18	12 04	23 01	10 36	13	
5	04 23	04 28	12 04	23 49	11 25	14	

Figure 38.

The value of 366°61.8′ was selected in this example because it demonstrates two points. The first is that any time the minutes of an angle exceed 60, the 60 minutes (60′) constitute an additional degree and the angle should be restated to reflect this fact. Thus, 366°61.8′ should be restated as 367°01.8′.

The second point is that the Greenwich hour angle (GHA) can never have a value greater than 360°. After all, a complete circle only contains 360°. Thus, 367°01.8′ should correctly be stated as 07°01.8′. In other words, whenever GHA exceeds 360° you must subtract 360° from the GHA. See Greenwich time, pages 10-12 above. This fact is also apparent from observing the GHA as shown in the table in the *Nautical Almanac*; as GHA approaches 360°, it suddenly drops back to 13° or so. Never is it shown over 360°.

We have now established the GHA of the sun as 07°01.8′ at the time of observation.

Declination
(See also
Figures 12–13)

The declination for 1200 is shown as N 22°52.9′. Since our shot was actually made not at 1200 but at $12^h32^m25^s$, we should, for greater accuracy, interpolate for this additional 32 minutes of time. This can be done by eye as the declination changes so slowly as compared to the GHA. Examination of the Dec. column for July 4, 1982, will show that the change in Dec. from hour to hour averages about .2 and that the value of the Dec. is *decreasing* as the day goes on. Since we have approximately 30 minutes of time in our example, it would be appropriate to subtract .1 from the 1200^h Dec. to account for this additional time.

Therefore, we could note the sun's declination as of the time of our shot as N 22°52.8′.

Common sense will dictate the amount of interpolation required. If your shot was only a few minutes after the hour, you may ignore any correction of declination beyond the even hour. On the other hand, if the time of your shot is approaching the next half hour or hour, it would be well to recognize this with an appropriate correction. Although the hourly change in this example in July is small, only .2 of a minute, in certain months the change can amount to as much as one whole degree in just one hour.

Having obtained the basic GHA and declination from the *Almanac*, it is now necessary to apply the other corrections to your HS, just as was done with the noon sight, in order to arrive at your height observed (HO). See pages 17 through 23. First correct your HS for index error and dip. This yields your App. Alt., apparent altitude. Next, obtain from the Altitude Correction Tables, inside the front cover of the *Almanac* (reprinted herein as Appendix 4) the correction for the appropriate month and apparent altitude. This value is then applied (either added or subtracted, according to the sign given) to the apparent altitude, and you have obtained your HO, or height observed.

In our example, let us assume a height of eye of 10 feet and an index error of +2′. First, we enter the table for the dip correction. This yields a value of –3.1′ which is combined with the index error of –2′ for a total correction of –5.1′. Remember that if your sextant shows an error 2′ *above* zero you must *subtract* 2′ as a correction to account for the fact that your sextant is reading too high.

Thus:	HS	$30°13'$
	IE	$-2'$
	Dip	$-3.1'$
	App. Alt.	$30°7.9'$
	Alt. Corr.	$14.4'$
	HO =	$30°22.3'$

Using the Work Form

The use of an organized work form is highly desirable as it helps the navigator keep from omitting a step under the pressure of a pitching table and bleary eyes. Further, it will enable you to visualize each step and how it ties to the next. The form shown will help in keeping track of the bits and pieces of information you are developing so that you will obtain an LOP in the shortest possible time. (For blank form, see Appendix 8.)

Sun Sight
Work Form

Date: July 4, 1982   DR. Lat: N 37° 15'   DR. Long: W 75° 22'

Watch Time: 08 32 25   Height of Eye 10'   HS: 30° 13'   IE: -2'

Watch Error 00

Step 1

HS	30° 13'
IE (+ or −)	−2'
DIP (−)	−3.1'
App. Alt.	30° 7.9'
Alt. Corr. =	+14.4'
HO =	30° 22.3'
(HO = App. Alt. ± Alt. Corr.)	

Step 2

Watch Time	=	08 32 25
Watch Error ±	=	00
Zone Time	=	08 32 25
Zone Description	=	+4
GMT =		12 32 25

*(For daylight-savings time subtract one hour from Z D)

```
   360
 + 01 01.8              38.0
 ────────              -22.3
  367° 01.8           ────────
  -75° 01.8            15.7
         ┌─────┐
         │ 59. │
         └─────┘
```

Step 3 (From Nautical Almanac)

GHA—Hours	358° 55.5'
GHA—M&S	8° 06.3'
Total GHA (± 360°)	7° 01.8'
Asmd. Long.	(W 75° 01.8')
LHA*	292°
Asmd. Lat.	N 37°
Decl. N or S	N 22° 52.8'

Step 4 (From HO 249)

Z	84°
ZN	(84°)
d(+ or −)	+ 31'
Tab HC	30° 11'
Corr. for d	+ 27'
HC	30° 38'
HO	30° 22.3'
Intercept, T or A	A 15.7'

*Whole degree only
LHA = GHa + E long, − W long.

Figure 39.

57

Let us now work through our example sight, placing the developing information in the proper slots in the work form.

Work Form Steps 1, 2, and 3

Steps 1, 2, and 3 follow easily along from the discussions above.

Step 4 introduces some new concepts.

First, we enter the *Almanac* with the whole hour GMT (here, 1200). We next take the whole hour GHA from the adjacent column, 358°55.5′, and enter this value on the form by GHA — Hours. Now the value for 32^m25^s, the minutes and seconds of correction, are obtained from the Increments and Corrections table in the back of the *Almanac* (Appendix 2 hereof) and entered on the next line of the form: 8°06.3′. The two values are then added together and we have the total GHA: 366°61.8′. This is restated as 07°01.8′, as discussed on page 55 above.

Note here that beside the total GHA there are parentheses showing (–360°). This is to remind you that if the total GHA exceeds 360°, you must subtract 360° from it.

Local Hour Angle

On pages 34–36 we considered the Greenwich hour angle and learned that the GHA is really the angular distance which the sun has traveled west from Greenwich.

If we were to consider our own location on planet earth and note the angle between our location and the sun, always measuring westward, we would obtain another useful angle, the local hour angle (LHA).

The LHA can easily be established by adding one's easterly longitude or by subtracting one's westerly longitude from the Greenwich hour angle.

Thus, if our DR is 75°22' West, we simply subtract this from the GHA of 07°01.8'. Since we cannot subtract a larger number from a smaller one, however, we add 360° to the GHA, just so the subtraction will work. (Obviously, when you add 360° to any point on a circle, you end up just where you started, so no change is really made.)

Note that the LHA of 292° is the angular distance from the assumed position to the sun's GP going in the direction of the sun's travel. The LHA should never be measured backwards, or contrary to the direction of the sun's travel since it is 0 when the sun is overhead and increases continuously until is is overhead again.

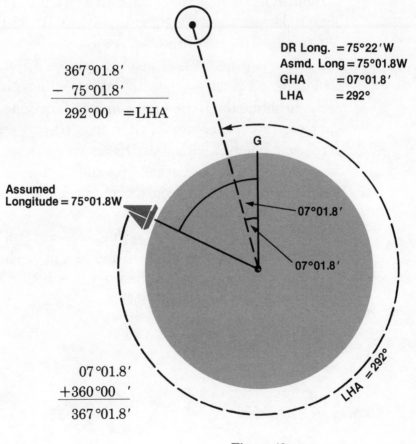

367°01.8'
− 75°01.8'
─────────
292°00   =LHA

DR Long. = 75°22'W
Asmd. Long = 75°01.8W
GHA   = 07°01.8'
LHA   = 292°

Assumed
Longitude = 75°01.8W

07°01.8'

07°01.8'

LHA = 292°

07°01.8'
+360°00   '
─────────
367°01.8'

Figure 40.

Assumed Longitude

You will note that the 4th space down in Step 3 of the work form calls for assumed longitude (Asmd. Long.). Note also that the DR longitude was W75°22', but that we used an *assumed longitude* of W75°01.8', slightly different from the DR. This is done intentionally so that the computation will work more easily. We use the correct *degrees* of longitude, but then arbitrarily use *minutes* which will cancel out the minutes of the GHA. In this way, the LHA will always be a whole degree; no minutes will appear in it to complicate matters. The tables we will be using only give values for the whole degrees; later we will interpolate for the missing minutes.

Remember that in West longitude, we use an assumed longitude with the same minutes as the GHA so that when the two are *subtracted* they will cancel out the minutes and leave a whole number of degrees for the LHA. In East longitude, however, we *add* the assumed longitude to the GHA to obtain the LHA so that in East longitude you must first subtract the minutes of GHA from 60' and use that difference in minutes with your DR degrees as your assumed longitude. In this way, when you add your assumed East longitude to the GHA, you will obtain a whole number of degrees for the local hour angle.

Making the LHA reflect whole degrees only will in no way impair the accuracy of our LOP, for, as will be shown later, the *assumed* longitude will be used in establishing an *assumed position* for plotting purposes.

Assumed Latitude
We will next enter on the form an *assumed latitude*. This will be a latitude with *degrees* only and *no minutes*. Here again, since we could be anywhere, we select a latitude which will make the rest of the computation easy and will not adversely affect the outcome. Thus, we assume and enter 37°, the whole degree nearest to our DR latitude.

The last blank on the work form to this point is for the declination. This is simply taken directly from the *Almanac*, and interpolated, if deemed necessary, for the passage of extra time from the whole hour as shown.

Chapter Seven:
Pub. No. 249
Sight Reduction Tables

We can now lay the *Nautical Almanac* aside and enter Pub. No. 249. This set of publications comes in three volumes: No. 1 — Tables for Selected Stars; No. 2 — Sun, Moon and Planets for Latitudes 0–40°; and No. 3 — Sun, Moon and Planets for Latitudes 39°–89°. Volumes No. 2 and 3 are valid perpetually, Volume 1 is revised every 5 years.

Since we are here concerned only with sun sights we need only volumes two and three of Pub. No. 249. With these two volumes and the current year's *Nautical Almanac*, you will have all the information you need to find your position anywhere on earth whenever observations of the sun are possible.

Since our assumed latitude is 37° N we will be concerned for now with Volume No. 2 only, since it deals with latitudes 0–40°.

Open Volume No. 2 and turn the pages until the numbers in the upper corners show Lat. 37°. Now turn the book so that the pages open away from you and you can read the tables easily.

From the information obtained from the *Nautical Almanac*, we have noted the sun's declination for the time of our observation to be N 22°52.8′. The N before the numbers, of course, indicates that the sun's declination is north of the equator, as it will always be during the summer months in the Northern Hemisphere. The reverse, of course, is true in the winter months when the sun has moved south, below the equator. See Figures 12 and 13.

Look now along the top and bottom of each page and you will see a notation that the declination is either the *same* as latitude or that declination is *contrary* to latitude.

Since we are in the *Northern* Hemisphere in our example and since the sun's declination is also *north* we must seek and use those pages that are labeled "Declination *Same Name* as Latitude."

Note also that following the word "declination" there are parentheses in which is indicated which declinations are shown on that page. Since the declination we are concerned with is 22° (whole degrees), we turn the pages until we find the one which indicates that declinations (15–29) are to be found on it, with Declination *Same Name* as Latitude.

Look now at the work form to determine again your LHA, in our example, 292°. On either side of the page there are columns indicating LHA. Look down the LHA numbers until you see 292, then read across to the 22° column. In our case the 292° is in the right-hand side, near the bottom. Reading down the 22° column to its intersection with the 292° LHA, the following information is obtained:

HC	d	Z
30°11′	+31′	84°

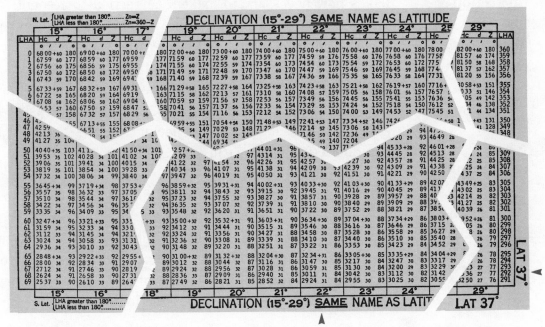

This page of *HO 249* is reproduced in full in Appendix 5 hereof.

Figure 41.

At this point this information should be entered on the work form in the appropriate spaces.

First: Enter the tabulated HC value just as it appears in Pub. No. 249 by the space marked "Tab HC": 30°11′.

Second: Enter the value of "d" as shown in the table just over this with either a + or – sign, such sign being indicated at the top of each block of 5 values in the table: +31′. ("d" is a factor used in obtaining the value of the missing minutes of declination.)

Third: Enter the value shown for "Z" in its appropriate space: 84°.

Things begin to get exciting at this point. Although certain corrections still must be applied to the tabulated HC, we can at least glance down at our HO to see if we are going to end up in the right ocean! Happily, we see we will, as our HO was 30°22.3′ and the uncorrected HC is 30°11′. It is the difference between the HO and the HC, called the *intercept*, which will form the basis of our LOP.

Correction for Minutes of Declination

It is now necessary to make a correction to account for the fact that we omitted the *minutes* of declination when we entered the Pub. No. 249 table. The Dec. as shown on the work form is 22°52.9′. We must correct our tabulated HC for the additional 52.8′ we ignored earlier. For practical purposes, we will round 52.8′ to 53′ as it will be easier to work with and changes in the declination are so slow that rounding off is not hurtful as respects accuracy.

On the last white page in Pub. No. 249 is found Table 5, "Correction to Tabulated Altitude for Minutes of Declination." This table is reproduced as Appendix 6 hereof. On each side of the table there is a column for values of "d." Across the top these are numbered 1–60. Read across the top numbers to 53 (our missing minutes of declination) and down the side to 31, the tabulated correction for "d."

Where the horizontal and vertical columns intersect, we read the correction for "d" as +27. This value should be entered on the work form beside "corr. for d" with the proper sign + or - shown. In the instant case, it is +27′. Do not forget that "d" can be either + or - and that the appropriate sign is given at the top of each group of 5 values.

The "corr. for d" is now added or subtracted, according to its sign, to or from the Tab. HC value of 30°11′ and we have the HC.

$$30°11′$$
$$\underline{+27′}$$
$$30°38′ = HC$$

This is entered on the work form as the HC.

Determining the Intercept

The last step prior to the actual plotting of the LOP is to subtract the HC from the HO. Either of these values may be larger or smaller than the other; it does not matter. Whichever is smaller, however, is subtracted from the larger and the resulting difference is called the intercept. In our example this is 15.7′ and should be entered in the last space on the form. This difference in minutes represents the distance in nautical miles of our actual position from our assumed position.

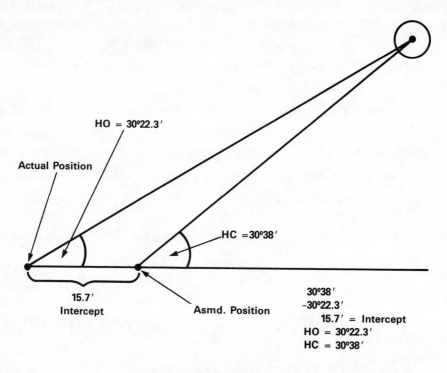

HO = 30°22.3'

Actual Position

HC =30°38'

15.7'
Intercept

Asmd. Position

30°38'
-30°22.3'
        15.7' = Intercept
HO = 30°22.3'
HC = 30°38'

Figure 42.

<table>
<tr><td>**Azimuth**</td><td>

The last bit of information we obtained from Pub. No. 249 was our "Z." Z relates to azimuth, which is the true bearing of the sun from our assumed position. "Zn" is the actual azimuth. Sometimes Z equals Zn, and sometimes it is necessary to add or subtract 180° or 360° to or from Z to obtain Zn.

Fortunately, determining Zn from Z is accomplished quite simply. It is only necessary to look in the upper or lower left corner of the page (depending on your latitude) in Pub. No. 249 where the declination and LHA were found to see what if anything is required to obtain Zn from Z.

Since in the present situation we are in a northern latitude, the upper left legend tells us that if LHA is greater than 180° (here 292°), Zn=Z. This is our situation, and so 84° is entered on the work form as the Zn.
</td></tr>
</table>

Azimuth is always established as a *true* bearing, never a *magnetic* bearing. Therefore, always be sure you are using the *true* compass rose when plotting the azimuth.

Homoto, the Japanese Navigator

There is a memory aid which will help in determining whether we should measure *towards* or *away* from the azimuth bearing from our assumed position. This is Homoto, the name of a mythical Japanese navigator. The letters actually stand for "HO is *more, towards*." The corollary, of course, being "if HO is *less, away*." In our example, HO is less so we will label the intercept "A," or *away* from the azimuth bearing. You can, of course, always simply draw a sketch to remind you of whether you should apply your intercept *towards* or *away* from the Zn bearing.

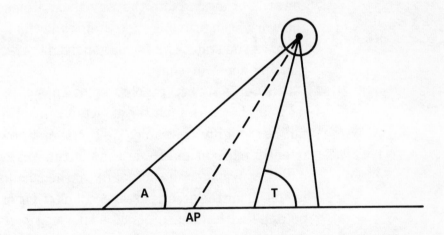

Figure 43.

Chapter Eight: Plotting the LOP

At this point, you are ready to plot on your chart or plotting sheet the information now formalized on your work form and thus obtain an LOP.

You will need:
1. Sharp pencil
2. Parallel rules
3. Dividers
4. A right angle or protractor

Just as an aid to help in plotting you may find it useful to draw a freehand ring around the items on the work form that you will need to use. These will include your assumed latitude, longitude, ZN and the *intercept* marked T or A, towards or away. Note the way these items are circled on the work form. Circling these items will simply help you spot them as you transfer information from the work form to the plotting sheet or chart.

It is perfectly acceptable to plot lines of position directly on your chart. As a matter of fact, when approaching a landfall, it is highly desirable. Well offshore, however, it is frequently more useful to do the plotting on a plotting sheet and then transfer the actual fix to the small scale chart being used, for the ocean passage. All plotting, of course, should be done with a sharp pencil — never a pen.

Plotting Sheets

Plotting sheets are available from most marine chandleries constructed for the particular latitudes in which you expect to be sailing. There are also universal plotting sheets which will serve for all the mid latitudes by measuring against a graduated scale which represents the converging meridians of longitude. These are very useful indeed and are of a smaller, more easily handled size. One of these is displayed as Appendix 7 hereof.

When plotting an LOP using a plotting sheet, it is practical to use the coordinates of your assumed position as the two principle lines which cross the center of the true compass rose. See plotting sheets for problems 1, 2, and 3 for examples. This makes plotting the ZN quite simple as all that is required is to lay your parallel rule through the intersection of the two lines and thence to the appropriate degrees for the ZN on the true rose. Now draw a line from the ZN through the AP at the center of the plotting sheet and on off towards the other side of the rose.

When plotting on a chart, however, the situation will be slightly different. Here you will find the horizontal or latitude line that represents your assumed latitude. With your dividers, measure along the scale at either the top or bottom of the chart the required degrees or minutes of longitude using whatever meridian of longitude is the closest to your assumed position and adding or subtracting the required minutes to or from that reference line.

When you have determined your assumed position on the chart, place a small dot there and label it "AP" for "assumed position."

Next, measure the length of the intercept, here 15.7 nautical miles, with the dividers and note whether it is marked "T" or "A," towards or away. If the intercept is "T," place one point of the dividers at the AP and measure the required intercept distance in the direction of the ZN. If it is marked "A," as in this example, then measure the intercept from the AP along the line *away* from the direction of the ZN, as determined from the true compass rose.

Remember that each minute of the intercept is one nautical mile, and you must therefore use the scale at either side of the chart to measure the intercept length. On these scales, each minute equals one nautical mile. Never use the scale at the top or bottom of the chart for this purpose as it will not be accurate since the top and bottom scales measure the meridians of longitude which do not represent nautical miles.

As stated above, the one time you will use the scale at the top or bottom of the chart will be in establishing the AP, i.e., in measuring east or west for minutes of longitude from a convenient labelled meridian.

Once the intercept has been measured from the AP, a dot should be placed there on the ZN line and a line drawn at this point at a right angle to the ZN line.

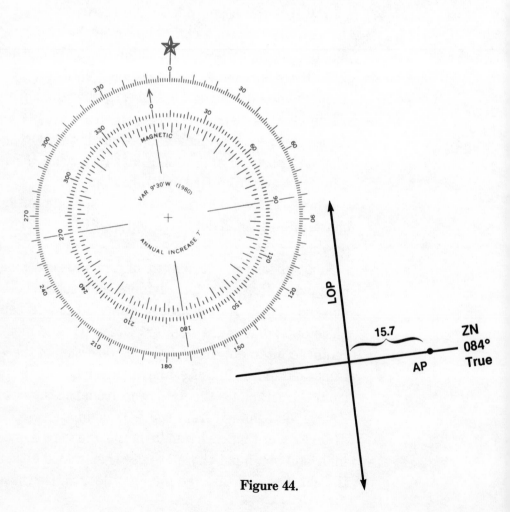

Figure 44.

There are many ways to establish a right angle from the ZN line. One of the easiest is to have a transparent right triangle such as draftsmen use; another is to use a protractor. In a pinch, however, almost any square cut sheet of paper or cardboard will do.

The LOP should be labelled and the time of the observation noted by it as well. At that time you were somewhere on that line. A point is usually drawn at each end of the LOP to indicate that it extends beyond what is shown on the chart.

In fact, the LOP thus constructed is really just a piece of an arc of a circle of equal altitude. It is permissible to show the LOP as a straight line for practical purposes because the circle of equal altitude is so vast that a piece of its arc as small as we are considering would not show any curve at all.

The Running Fix

There are times when a single LOP will be most helpful. Frequently, such as on approaching a coast, it may be possible to cross a celestial LOP with a radio direction finder signal and obtain an accurate fix. When such opportunities exist, they should be seized, but there will also be many times far from civilization when no LOP other than a celestial one can be had. On these occasions, it is possible to obtain a good, accurate fix using just sun lines.

The procedure here is to shoot the sun early in the morning and plot this LOP on the chart or plotting sheet. As we learned from the flagpole examples above, what we obtain from an observation of the sun is an LOP perpendicular to a line from our position towards the sun.

If we were to stay in the same place and shoot the sun several hours later, we would have a second LOP which would intersect the first; we would have a fix.

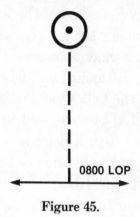

0800 LOP

Figure 45.

We could then continue to shoot sun lines all day and all should intersect at our position, if we remained completely stationary.

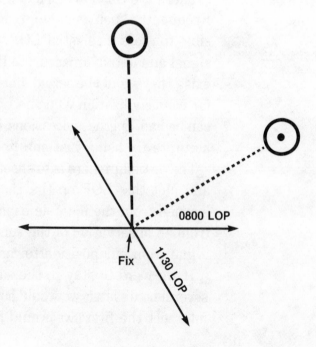

0800 LOP

1130 LOP

Fix

Figure 46.

72

It is not practical, however, to remain in one position if one is to get anywhere and so we must move, or *advance*, our earlier LOP to allow for the movement of our boat between the shots.

Advancing the LOP is quite simple. From any point on the first LOP, draw a line that represents your course. Label this line by placing a large "C" over it followed by the course, in a 3-digit number. Under the line place a large "S" followed by your estimated, or determined, speed.

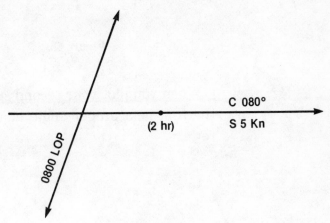

Figure 47.

If 2 hours have elapsed between your first shot and your next, and you are averaging 05 knots, measure with your dividers down the course line from your original position on the LOP the 10 nautical miles you have travelled in those 2 hours and place a dot there on the course line.

Now with the parallel rule, advance the first sun line down the course line to the dot and draw a line through it parallel to the first sun line. This should be labeled "Advanced 0800 LOP."

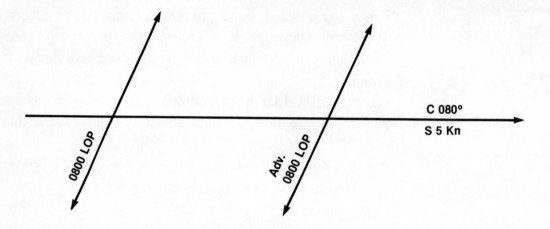

Figure 48.

When you plot your second sun line now, it will intersect the first at a proper angle and a reliable fix will be your reward.

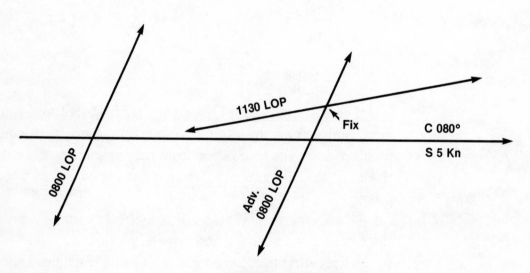

Figure 49.

This process can be repeated all during the daylight hours. The more lines that are plotted and advanced, the greater the accuracy of the fix will be as there is always the possibility of some error in the measurement or estimate of distance travelled. Causes for this might be unknown current, more leeway than expected, undetected compass error, or (Heaven forbid!) an inattentive helmsman.

As stated above, a course line should be labeled with a large "C" above the line followed by the course expressed in three digits, e.g., 085°. Although big ships always label the *true* course, it is perfectly acceptable on small vessels to indicate a magnetic course provided the abbreviation "mag." is placed after the three digits, i.e., "085° mag."

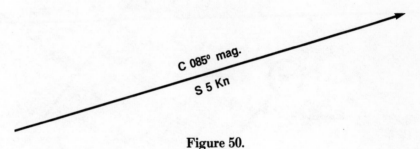

Figure 50.

When plotting a second LOP on a plotting sheet, the first line is advanced in the usual way. The new assumed position, however, will not be located at the intersection of the latitude and longitude lines at the center of the compass rose but must be located using the dividers and the longitude scale at the side of the plotting sheet and establishing the new assumed longitude by measuring from the original one.

The horizontal line representing latitude will remain the same, naturally, since the boat will not have moved from 1 whole degree of latitude to another within the course of only a few hours.

Experience will soon show the navigator that when he is on course he will obtain an LOP indicating his speed if he shoots the sun when it is either dead ahead or dead astern of his vessel. Likewise, a shot when the sun is on either beam will yield a line parallel to his course.

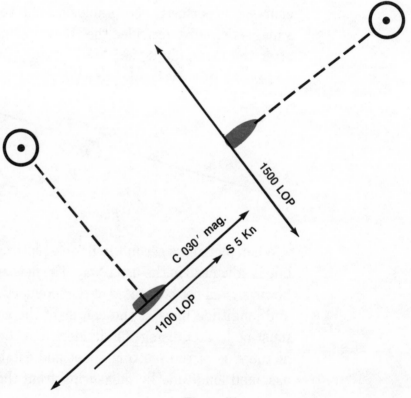

Figure 51

In the early morning or late afternoon when the sun is low in the eastern or western sky, a sextant shot will yield a sun line running pretty much north/south, useful for longitude.

Towards midday, when the sun is high in the sky, a shot will yield an east/west line, useful for latitude.

Problems

Thus, if you are crossing an ocean on an east/west heading, midday shots will tell you whether or not you are on your course; early morning and late afternoon shots will tell you how far along that course you have travelled.

Solve the following for intercept and plot sun lines:

1. July 4, 1982; Watch time of observation 07 32 17; Watch error +12″; Height of Eye 10 feet; DR Lat. N 36°51′; Long. W 71°20′; Instrument error –03′; HS 21°08′, lower limb.

2. January 2, 1982; Watch time of observation 15 23 44; Watch error – 05′; Height of Eye 10 feet; DR Lat. S 17°15′; Long. W 148°47′; Instrument error +04′; HS=41°27.2′, lower limb.

3. January 3, 1982; Watch time of observation 13 05 26; Watch error +06″; Height of Eye 10 feet; DR Lat. S 32°43′; DR Long. E 16°22′; Instrument error +06′; HS 72°41′, lower limb.

(Answers on pages 106–111).

All of the information from the *Nautical Almanac* and Pub. No. 249, required to solve the above problems, will be found in Appendixes 1 through 6.

Conclusion

As has been shown, there is really nothing difficult or mysterious in obtaining an accurate fix from observations of the sun taken several hours apart. Observations taken too close together, however, will yield LOP's which cross at angles which are too acute to be accurate. Shots taken at three to four hour intervals will yield LOP's that intersect at 45° or more and these will be most practical in establishing fixes.

Practice with your sextant, watch, tables, and work forms until you have gained confidence in your ability to drop a sun line squarely through your position standing on a beach or on your boat in a bay or estuary. Then when you seriously contemplate that blue water passage, you will not be terrified of the magic black boxes winking out with salt air and corrosion.

The ability to navigate by the sun will add a new dimension to your cruises and new zest to your adventure.

Good luck and happy landfalls!

APPENDICES

SUN SIGHT NAVIGATION

APPENDIX 1

CONVERSION OF ARC TO TIME

0°–59°	h m	60°–119°	h m	120°–179°	h m	180°–239°	h m	240°–299°	h m	300°–359°	h m	′	0′·00 m s	0′·25 m s	0′·50 m s	0′·75 m s
0	0 00	60	4 00	120	8 00	180	12 00	240	16 00	300	20 00	0	0 00	0 01	0 02	0 03
1	0 04	61	4 04	121	8 04	181	12 04	241	16 04	301	20 04	1	0 04	0 05	0 06	0 07
2	0 08	62	4 08	122	8 08	182	12 08	242	16 08	302	20 08	2	0 08	0 09	0 10	0 11
3	0 12	63	4 12	123	8 12	183	12 12	243	16 12	303	20 12	3	0 12	0 13	0 14	0 15
4	0 16	64	4 16	124	8 16	184	12 16	244	16 16	304	20 16	4	0 16	0 17	0 18	0 19
5	0 20	65	4 20	125	8 20	185	12 20	245	16 20	305	20 20	5	0 20	0 21	0 22	0 23
6	0 24	66	4 24	126	8 24	186	12 24	246	16 24	306	20 24	6	0 24	0 25	0 26	0 27
7	0 28	67	4 28	127	8 28	187	12 28	247	16 28	307	20 28	7	0 28	0 29	0 30	0 31
8	0 32	68	4 32	128	8 32	188	12 32	248	16 32	308	20 32	8	0 32	0 33	0 34	0 35
9	0 36	69	4 36	129	8 36	189	12 36	249	16 36	309	20 36	9	0 36	0 37	0 38	0 39
10	0 40	70	4 40	130	8 40	190	12 40	250	16 40	310	20 40	10	0 40	0 41	0 42	0 43
11	0 44	71	4 44	131	8 44	191	12 44	251	16 44	311	20 44	11	0 44	0 45	0 46	0 47
12	0 48	72	4 48	132	8 48	192	12 48	252	16 48	312	20 48	12	0 48	0 49	0 50	0 51
13	0 52	73	4 52	133	8 52	193	12 52	253	16 52	313	20 52	13	0 52	0 53	0 54	0 55
14	0 56	74	4 56	134	8 56	194	12 56	254	16 56	314	20 56	14	0 56	0 57	0 58	0 59
15	1 00	75	5 00	135	9 00	195	13 00	255	17 00	315	21 00	15	1 00	1 01	1 02	1 03
16	1 04	76	5 04	136	9 04	196	13 04	256	17 04	316	21 04	16	1 04	1 05	1 06	1 07
17	1 08	77	5 08	137	9 08	197	13 08	257	17 08	317	21 08	17	1 08	1 09	1 10	1 11
18	1 12	78	5 12	138	9 12	198	13 12	258	17 12	318	21 12	18	1 12	1 13	1 14	1 15
19	1 16	79	5 16	139	9 16	199	13 16	259	17 16	319	21 16	19	1 16	1 17	1 18	1 19
20	1 20	80	5 20	140	9 20	200	13 20	260	17 20	320	21 20	20	1 20	1 21	1 22	1 23
21	1 24	81	5 24	141	9 24	201	13 24	261	17 24	321	21 24	21	1 24	1 25	1 26	1 27
22	1 28	82	5 28	142	9 28	202	13 28	262	17 28	322	21 28	22	1 28	1 29	1 30	1 31
23	1 32	83	5 32	143	9 32	203	13 32	263	17 32	323	21 32	23	1 32	1 33	1 34	1 35
24	1 36	84	5 36	144	9 36	204	13 36	264	17 36	324	21 36	24	1 36	1 37	1 38	1 39
25	1 40	85	5 40	145	9 40	205	13 40	265	17 40	325	21 40	25	1 40	1 41	1 42	1 43
26	1 44	86	5 44	146	9 44	206	13 44	266	17 44	326	21 44	26	1 44	1 45	1 46	1 47
27	1 48	87	5 48	147	9 48	207	13 48	267	17 48	327	21 48	27	1 48	1 49	1 50	1 51
28	1 52	88	5 52	148	9 52	208	13 52	268	17 52	328	21 52	28	1 52	1 53	1 54	1 55
29	1 56	89	5 56	149	9 56	209	13 56	269	17 56	329	21 56	29	1 56	1 57	1 58	1 59
30	2 00	90	6 00	150	10 00	210	14 00	270	18 00	330	22 00	30	2 00	2 01	2 02	2 03
31	2 04	91	6 04	151	10 04	211	14 04	271	18 04	331	22 04	31	2 04	2 05	2 06	2 07
32	2 08	92	6 08	152	10 08	212	14 08	272	18 08	332	22 08	32	2 08	2 09	2 10	2 11
33	2 12	93	6 12	153	10 12	213	14 12	273	18 12	333	22 12	33	2 12	2 13	2 14	2 15
34	2 16	94	6 16	154	10 16	214	14 16	274	18 16	334	22 16	34	2 16	2 17	2 18	2 19
35	2 20	95	6 20	155	10 20	215	14 20	275	18 20	335	22 20	35	2 20	2 21	2 22	2 23
36	2 24	96	6 24	156	10 24	216	14 24	276	18 24	336	22 24	36	2 24	2 25	2 26	2 27
37	2 28	97	6 28	157	10 28	217	14 28	277	18 28	337	22 28	37	2 28	2 29	2 30	2 31
38	2 32	98	6 32	158	10 32	218	14 32	278	18 32	338	22 32	38	2 32	2 33	2 34	2 35
39	2 36	99	6 36	159	10 36	219	14 36	279	18 36	339	22 36	39	2 36	2 37	2 38	2 39
40	2 40	100	6 40	160	10 40	220	14 40	280	18 40	340	22 40	40	2 40	2 41	2 42	2 43
41	2 44	101	6 44	161	10 44	221	14 44	281	18 44	341	22 44	41	2 44	2 45	2 46	2 47
42	2 48	102	6 48	162	10 48	222	14 48	282	18 48	342	22 48	42	2 48	2 49	2 50	2 51
43	2 52	103	6 52	163	10 52	223	14 52	283	18 52	343	22 52	43	2 52	2 53	2 54	2 55
44	2 56	104	6 56	164	10 56	224	14 56	284	18 56	344	22 56	44	2 56	2 57	2 58	2 59
45	3 00	105	7 00	165	11 00	225	15 00	285	19 00	345	23 00	45	3 00	3 01	3 02	3 03
46	3 04	106	7 04	166	11 04	226	15 04	286	19 04	346	23 04	46	3 04	3 05	3 06	3 07
47	3 08	107	7 08	167	11 08	227	15 08	287	19 08	347	23 08	47	3 08	3 09	3 10	3 11
48	3 12	108	7 12	168	11 12	228	15 12	288	19 12	348	23 12	48	3 12	3 13	3 14	3 15
49	3 16	109	7 16	169	11 16	229	15 16	289	19 16	349	23 16	49	3 16	3 17	3 18	3 19
50	3 20	110	7 20	170	11 20	230	15 20	290	19 20	350	23 20	50	3 20	3 21	3 22	3 23
51	3 24	111	7 24	171	11 24	231	15 24	291	19 24	351	23 24	51	3 24	3 25	3 26	3 27
52	3 28	112	7 28	172	11 28	232	15 28	292	19 28	352	23 28	52	3 28	3 29	3 30	3 31
53	3 32	113	7 32	173	11 32	233	15 32	293	19 32	353	23 32	53	3 32	3 33	3 34	3 35
54	3 36	114	7 36	174	11 36	234	15 36	294	19 36	354	23 36	54	3 36	3 37	3 38	3 39
55	3 40	115	7 40	175	11 40	235	15 40	295	19 40	355	23 40	55	3 40	3 41	3 42	3 43
56	3 44	116	7 44	176	11 44	236	15 44	296	19 44	356	23 44	56	3 44	3 45	3 46	3 47
57	3 48	117	7 48	177	11 48	237	15 48	297	19 48	357	23 48	57	3 48	3 49	3 50	3 51
58	3 52	118	7 52	178	11 52	238	15 52	298	19 52	358	23 52	58	3 52	3 53	3 54	3 55
59	3 56	119	7 56	179	11 56	239	15 56	299	19 56	359	23 56	59	3 56	3 57	3·58	3 59

The above table is for converting expressions in arc to their equivalent in time ; its main use in this Almanac is for the conversion of longitude for application to L.M.T. (*added* if *west*, *subtracted* if *east*) to give G.M.T. or vice versa, particularly in the case of sunrise, sunset, etc.

i

284-121 O – 81 – 18 : QL 3

INCREMENTS AND CORRECTIONS

2ᵐ   **3ᵐ**

2ᵐ

2 (s)	SUN PLANETS	ARIES	MOON	v or Corrⁿ d	v or Corrⁿ d	v or Corrⁿ d
00	0 30·0	0 30·1	0 28·6	0·0 0·0	6·0 0·3	12·0 0·5
01	0 30·3	0 30·3	0 28·9	0·1 0·0	6·1 0·3	12·1 0·5
02	0 30·5	0 30·6	0 29·1	0·2 0·0	6·2 0·3	12·2 0·5
03	0 30·8	0 30·8	0 29·3	0·3 0·0	6·3 0·3	12·3 0·5
04	0 31·0	0 31·1	0 29·6	0·4 0·0	6·4 0·3	12·4 0·5
05	0 31·3	0 31·3	0 29·8	0·5 0·0	6·5 0·3	12·5 0·5
06	0 31·5	0 31·6	0 30·1	0·6 0·0	6·6 0·3	12·6 0·5
07	0 31·8	0 31·8	0 30·3	0·7 0·0	6·7 0·3	12·7 0·5
08	0 32·0	0 32·1	0 30·5	0·8 0·0	6·8 0·3	12·8 0·5
09	0 32·3	0 32·3	0 30·8	0·9 0·0	6·9 0·3	12·9 0·5
10	0 32·5	0 32·6	0 31·0	1·0 0·0	7·0 0·3	13·0 0·5
11	0 32·8	0 32·8	0 31·3	1·1 0·0	7·1 0·3	13·1 0·5
12	0 33·0	0 33·1	0 31·5	1·2 0·1	7·2 0·3	13·2 0·6
13	0 33·3	0 33·3	0 31·7	1·3 0·1	7·3 0·3	13·3 0·6
14	0 33·5	0 33·6	0 32·0	1·4 0·1	7·4 0·3	13·4 0·6
15	0 33·8	0 33·8	0 32·2	1·5 0·1	7·5 0·3	13·5 0·6
16	0 34·0	0 34·1	0 32·5	1·6 0·1	7·6 0·3	13·6 0·6
17	0 34·3	0 34·3	0 32·7	1·7 0·1	7·7 0·3	13·7 0·6
18	0 34·5	0 34·6	0 32·9	1·8 0·1	7·8 0·3	13·8 0·6
19	0 34·8	0 34·8	0 33·2	1·9 0·1	7·9 0·3	13·9 0·6
20	0 35·0	0 35·1	0 33·4	2·0 0·1	8·0 0·3	14·0 0·6
21	0 35·3	0 35·3	0 33·6	2·1 0·1	8·1 0·3	14·1 0·6
22	0 35·5	0 35·6	0 33·9	2·2 0·1	8·2 0·3	14·2 0·6
23	0 35·8	0 35·8	0 34·1	2·3 0·1	8·3 0·3	14·3 0·6
24	0 36·0	0 36·1	0 34·4	2·4 0·1	8·4 0·4	14·4 0·6
25	0 36·3	0 36·3	0 34·6	2·5 0·1	8·5 0·4	14·5 0·6
26	0 36·5	0 36·6	0 34·8	2·6 0·1	8·6 0·4	14·6 0·6
27	0 36·8	0 36·9	0 35·1	2·7 0·1	8·7 0·4	14·7 0·6
28	0 37·0	0 37·1	0 35·3	2·8 0·1	8·8 0·4	14·8 0·6
29	0 37·3	0 37·4	0 35·6	2·9 0·1	8·9 0·4	14·9 0·6
30	0 37·5	0 37·6	0 35·8	3·0 0·1	9·0 0·4	15·0 0·6
31	0 37·8	0 37·9	0 36·0	3·1 0·1	9·1 0·4	15·1 0·6
32	0 38·0	0 38·1	0 36·3	3·2 0·1	9·2 0·4	15·2 0·6
33	0 38·3	0 38·4	0 36·5	3·3 0·1	9·3 0·4	15·3 0·6
34	0 38·5	0 38·6	0 36·7	3·4 0·1	9·4 0·4	15·4 0·6
35	0 38·8	0 38·9	0 37·0	3·5 0·1	9·5 0·4	15·5 0·6
36	0 39·0	0 39·1	0 37·2	3·6 0·2	9·6 0·4	15·6 0·7
37	0 39·3	0 39·4	0 37·5	3·7 0·2	9·7 0·4	15·7 0·7
38	0 39·5	0 39·6	0 37·7	3·8 0·2	9·8 0·4	15·8 0·7
39	0 39·8	0 39·9	0 37·9	3·9 0·2	9·9 0·4	15·9 0·7
40	0 40·0	0 40·1	0 38·2	4·0 0·2	10·0 0·4	16·0 0·7
41	0 40·3	0 40·4	0 38·4	4·1 0·2	10·1 0·4	16·1 0·7
42	0 40·5	0 40·6	0 38·7	4·2 0·2	10·2 0·4	16·2 0·7
43	0 40·8	0 40·9	0 38·9	4·3 0·2	10·3 0·4	16·3 0·7
44	0 41·0	0 41·1	0 39·1	4·4 0·2	10·4 0·4	16·4 0·7
45	0 41·3	0 41·4	0 39·4	4·5 0·2	10·5 0·4	16·5 0·7
46	0 41·5	0 41·6	0 39·6	4·6 0·2	10·6 0·4	16·6 0·7
47	0 41·8	0 41·9	0 39·8	4·7 0·2	10·7 0·4	16·7 0·7
48	0 42·0	0 42·1	0 40·1	4·8 0·2	10·8 0·5	16·8 0·7
49	0 42·3	0 42·4	0 40·3	4·9 0·2	10·9 0·5	16·9 0·7
50	0 42·5	0 42·6	0 40·6	5·0 0·2	11·0 0·5	17·0 0·7
51	0 42·8	0 42·9	0 40·8	5·1 0·2	11·1 0·5	17·1 0·7
52	0 43·0	0 43·1	0 41·0	5·2 0·2	11·2 0·5	17·2 0·7
53	0 43·3	0 43·4	0 41·3	5·3 0·2	11·3 0·5	17·3 0·7
54	0 43·5	0 43·6	0 41·5	5·4 0·2	11·4 0·5	17·4 0·7
55	0 43·8	0 43·9	0 41·8	5·5 0·2	11·5 0·5	17·5 0·7
56	0 44·0	0 44·1	0 42·0	5·6 0·2	11·6 0·5	17·6 0·7
57	0 44·3	0 44·4	0 42·2	5·7 0·2	11·7 0·5	17·7 0·7
58	0 44·5	0 44·6	0 42·5	5·8 0·2	11·8 0·5	17·8 0·7
59	0 44·8	0 44·9	0 42·7	5·9 0·2	11·9 0·5	17·9 0·7
60	0 45·0	0 45·1	0 43·0	6·0 0·3	12·0 0·5	18·0 0·8

3ᵐ

3 (s)	SUN PLANETS	ARIES	MOON	v or Corrⁿ d	v or Corrⁿ d	v or Corrⁿ d
00	0 45·0	0 45·1	0 43·0	0·0 0·0	6·0 0·4	12·0 0·7
01	0 45·3	0 45·4	0 43·2	0·1 0·0	6·1 0·4	12·1 0·7
02	0 45·5	0 45·6	0 43·4	0·2 0·0	6·2 0·4	12·2 0·7
03	0 45·8	0 45·9	0 43·7	0·3 0·0	6·3 0·4	12·3 0·7
04	0 46·0	0 46·1	0 43·9	0·4 0·0	6·4 0·4	12·4 0·7
05	0 46·3	0 46·4	0 44·1	0·5 0·0	6·5 0·4	12·5 0·7
06	0 46·5	0 46·6	0 44·4	0·6 0·0	6·6 0·4	12·6 0·7
07	0 46·8	0 46·9	0 44·6	0·7 0·0	6·7 0·4	12·7 0·7
08	0 47·0	0 47·1	0 44·9	0·8 0·0	6·8 0·4	12·8 0·7
09	0 47·3	0 47·4	0 45·1	0·9 0·1	6·9 0·4	12·9 0·8
10	0 47·5	0 47·6	0 45·3	1·0 0·1	7·0 0·4	13·0 0·8
11	0 47·8	0 47·9	0 45·6	1·1 0·1	7·1 0·4	13·1 0·8
12	0 48·0	0 48·1	0 45·8	1·2 0·1	7·2 0·4	13·2 0·8
13	0 48·3	0 48·4	0 46·1	1·3 0·1	7·3 0·4	13·3 0·8
14	0 48·5	0 48·6	0 46·3	1·4 0·1	7·4 0·4	13·4 0·8
15	0 48·8	0 48·9	0 46·5	1·5 0·1	7·5 0·4	13·5 0·8
16	0 49·0	0 49·1	0 46·8	1·6 0·1	7·6 0·4	13·6 0·8
17	0 49·3	0 49·4	0 47·0	1·7 0·1	7·7 0·4	13·7 0·8
18	0 49·5	0 49·6	0 47·2	1·8 0·1	7·8 0·5	13·8 0·8
19	0 49·8	0 49·9	0 47·5	1·9 0·1	7·9 0·5	13·9 0·8
20	0 50·0	0 50·1	0 47·7	2·0 0·1	8·0 0·5	14·0 0·8
21	0 50·3	0 50·4	0 48·0	2·1 0·1	8·1 0·5	14·1 0·8
22	0 50·5	0 50·6	0 48·2	2·2 0·1	8·2 0·5	14·2 0·8
23	0 50·8	0 50·9	0 48·4	2·3 0·1	8·3 0·5	14·3 0·8
24	0 51·0	0 51·1	0 48·7	2·4 0·1	8·4 0·5	14·4 0·8
25	0 51·3	0 51·4	0 48·9	2·5 0·1	8·5 0·5	14·5 0·8
26	0 51·5	0 51·6	0 49·2	2·6 0·2	8·6 0·5	14·6 0·9
27	0 51·8	0 51·9	0 49·4	2·7 0·2	8·7 0·5	14·7 0·9
28	0 52·0	0 52·1	0 49·6	2·8 0·2	8·8 0·5	14·8 0·9
29	0 52·3	0 52·4	0 49·9	2·9 0·2	8·9 0·5	14·9 0·9
30	0 52·5	0 52·6	0 50·1	3·0 0·2	9·0 0·5	15·0 0·9
31	0 52·8	0 52·9	0 50·3	3·1 0·2	9·1 0·5	15·1 0·9
32	0 53·0	0 53·1	0 50·6	3·2 0·2	9·2 0·5	15·2 0·9
33	0 53·3	0 53·4	0 50·8	3·3 0·2	9·3 0·5	15·3 0·9
34	0 53·5	0 53·6	0 51·1	3·4 0·2	9·4 0·5	15·4 0·9
35	0 53·8	0 53·9	0 51·3	3·5 0·2	9·5 0·6	15·5 0·9
36	0 54·0	0 54·1	0 51·5	3·6 0·2	9·6 0·6	15·6 0·9
37	0 54·3	0 54·4	0 51·8	3·7 0·2	9·7 0·6	15·7 0·9
38	0 54·5	0 54·6	0 52·0	3·8 0·2	9·8 0·6	15·8 0·9
39	0 54·8	0 54·9	0 52·3	3·9 0·2	9·9 0·6	15·9 0·9
40	0 55·0	0 55·2	0 52·5	4·0 0·2	10·0 0·6	16·0 0·9
41	0 55·3	0 55·4	0 52·7	4·1 0·2	10·1 0·6	16·1 0·9
42	0 55·5	0 55·7	0 53·0	4·2 0·2	10·2 0·6	16·2 0·9
43	0 55·8	0 55·9	0 53·2	4·3 0·3	10·3 0·6	16·3 1·0
44	0 56·0	0 56·2	0 53·4	4·4 0·3	10·4 0·6	16·4 1·0
45	0 56·3	0 56·4	0 53·7	4·5 0·3	10·5 0·6	16·5 1·0
46	0 56·5	0 56·7	0 53·9	4·6 0·3	10·6 0·6	16·6 1·0
47	0 56·8	0 56·9	0 54·2	4·7 0·3	10·7 0·6	16·7 1·0
48	0 57·0	0 57·2	0 54·4	4·8 0·3	10·8 0·6	16·8 1·0
49	0 57·3	0 57·4	0 54·6	4·9 0·3	10·9 0·6	16·9 1·0
50	0 57·5	0 57·7	0 54·9	5·0 0·3	11·0 0·6	17·0 1·0
51	0 57·8	0 57·9	0 55·1	5·1 0·3	11·1 0·6	17·1 1·0
52	0 58·0	0 58·2	0 55·4	5·2 0·3	11·2 0·7	17·2 1·0
53	0 58·3	0 58·4	0 55·6	5·3 0·3	11·3 0·7	17·3 1·0
54	0 58·5	0 58·7	0 55·8	5·4 0·3	11·4 0·7	17·4 1·0
55	0 58·8	0 58·9	0 56·1	5·5 0·3	11·5 0·7	17·5 1·0
56	0 59·0	0 59·2	0 56·3	5·6 0·3	11·6 0·7	17·6 1·0
57	0 59·3	0 59·4	0 56·6	5·7 0·3	11·7 0·7	17·7 1·0
58	0 59·5	0 59·7	0 56·8	5·8 0·3	11·8 0·7	17·8 1·0
59	0 59·8	0 59·9	0 57·0	5·9 0·3	11·9 0·7	17·9 1·0
60	1 00·0	1 00·2	0 57·3	6·0 0·4	12·0 0·7	18·0 1·1

APPENDIX 2B

4^m   INCREMENTS AND CORRECTIONS   **5^m**

4^m

4 (s)	SUN PLANETS	ARIES	MOON	v or d / Corrn	v or d / Corrn	v or d / Corrn
00	1 00·0	1 00·2	0 57·3	0·0 0·0	6·0 0·5	12·0 0·9
01	1 00·3	1 00·4	0 57·5	0·1 0·0	6·1 0·5	12·1 0·9
02	1 00·5	1 00·7	0 57·7	0·2 0·0	6·2 0·5	12·2 0·9
03	1 00·8	1 00·9	0 58·0	0·3 0·0	6·3 0·5	12·3 0·9
04	1 01·0	1 01·2	0 58·2	0·4 0·0	6·4 0·5	12·4 0·9
05	1 01·3	1 01·4	0 58·5	0·5 0·0	6·5 0·5	12·5 0·9
06	1 01·5	1 01·7	0 58·7	0·6 0·0	6·6 0·5	12·6 0·9
07	1 01·8	1 01·9	0 58·9	0·7 0·1	6·7 0·5	12·7 1·0
08	1 02·0	1 02·2	0 59·2	0·8 0·1	6·8 0·5	12·8 1·0
09	1 02·3	1 02·4	0 59·4	0·9 0·1	6·9 0·5	12·9 1·0
10	1 02·5	1 02·7	0 59·7	1·0 0·1	7·0 0·5	13·0 1·0
11	1 02·8	1 02·9	0 59·9	1·1 0·1	7·1 0·5	13·1 1·0
12	1 03·0	1 03·2	1 00·1	1·2 0·1	7·2 0·5	13·2 1·0
13	1 03·3	1 03·4	1 00·4	1·3 0·1	7·3 0·5	13·3 1·0
14	1 03·5	1 03·7	1 00·6	1·4 0·1	7·4 0·6	13·4 1·0
15	1 03·8	1 03·9	1 00·8	1·5 0·1	7·5 0·6	13·5 1·0
16	1 04·0	1 04·2	1 01·1	1·6 0·1	7·6 0·6	13·6 1·0
17	1 04·3	1 04·4	1 01·3	1·7 0·1	7·7 0·6	13·7 1·0
18	1 04·5	1 04·7	1 01·6	1·8 0·1	7·8 0·6	13·8 1·0
19	1 04·8	1 04·9	1 01·8	1·9 0·1	7·9 0·6	13·9 1·0
20	1 05·0	1 05·2	1 02·0	2·0 0·2	8·0 0·6	14·0 1·1
21	1 05·3	1 05·4	1 02·3	2·1 0·2	8·1 0·6	14·1 1·1
22	1 05·5	1 05·7	1 02·5	2·2 0·2	8·2 0·6	14·2 1·1
23	1 05·8	1 05·9	1 02·8	2·3 0·2	8·3 0·6	14·3 1·1
24	1 06·0	1 06·2	1 03·0	2·4 0·2	8·4 0·6	14·4 1·1
25	1 06·3	1 06·4	1 03·2	2·5 0·2	8·5 0·6	14·5 1·1
26	1 06·5	1 06·7	1 03·5	2·6 0·2	8·6 0·6	14·6 1·1
27	1 06·8	1 06·9	1 03·7	2·7 0·2	8·7 0·7	14·7 1·1
28	1 07·0	1 07·2	1 03·9	2·8 0·2	8·8 0·7	14·8 1·1
29	1 07·3	1 07·4	1 04·2	2·9 0·2	8·9 0·7	14·9 1·1
30	1 07·5	1 07·7	1 04·4	3·0 0·2	9·0 0·7	15·0 1·1
31	1 07·8	1 07·9	1 04·7	3·1 0·2	9·1 0·7	15·1 1·1
32	1 08·0	1 08·2	1 04·9	3·2 0·2	9·2 0·7	15·2 1·1
33	1 08·3	1 08·4	1 05·1	3·3 0·2	9·3 0·7	15·3 1·1
34	1 08·5	1 08·7	1 05·4	3·4 0·3	9·4 0·7	15·4 1·2
35	1 08·8	1 08·9	1 05·6	3·5 0·3	9·5 0·7	15·5 1·2
36	1 09·0	1 09·2	1 05·9	3·6 0·3	9·6 0·7	15·6 1·2
37	1 09·3	1 09·4	1 06·1	3·7 0·3	9·7 0·7	15·7 1·2
38	1 09·5	1 09·7	1 06·3	3·8 0·3	9·8 0·7	15·8 1·2
39	1 09·8	1 09·9	1 06·6	3·9 0·3	9·9 0·7	15·9 1·2
40	1 10·0	1 10·2	1 06·8	4·0 0·3	10·0 0·8	16·0 1·2
41	1 10·3	1 10·4	1 07·0	4·1 0·3	10·1 0·8	16·1 1·2
42	1 10·5	1 10·7	1 07·3	4·2 0·3	10·2 0·8	16·2 1·2
43	1 10·8	1 10·9	1 07·5	4·3 0·3	10·3 0·8	16·3 1·2
44	1 11·0	1 11·2	1 07·8	4·4 0·3	10·4 0·8	16·4 1·2
45	1 11·3	1 11·4	1 08·0	4·5 0·3	10·5 0·8	16·5 1·2
46	1 11·5	1 11·7	1 08·2	4·6 0·3	10·6 0·8	16·6 1·2
47	1 11·8	1 11·9	1 08·5	4·7 0·4	10·7 0·8	16·7 1·3
48	1 12·0	1 12·2	1 08·7	4·8 0·4	10·8 0·8	16·8 1·3
49	1 12·3	1 12·4	1 09·0	4·9 0·4	10·9 0·8	16·9 1·3
50	1 12·5	1 12·7	1 09·2	5·0 0·4	11·0 0·8	17·0 1·3
51	1 12·8	1 12·9	1 09·4	5·1 0·4	11·1 0·8	17·1 1·3
52	1 13·0	1 13·2	1 09·7	5·2 0·4	11·2 0·8	17·2 1·3
53	1 13·3	1 13·5	1 09·9	5·3 0·4	11·3 0·8	17·3 1·3
54	1 13·5	1 13·7	1 10·2	5·4 0·4	11·4 0·9	17·4 1·3
55	1 13·8	1 14·0	1 10·4	5·5 0·4	11·5 0·9	17·5 1·3
56	1 14·0	1 14·2	1 10·6	5·6 0·4	11·6 0·9	17·6 1·3
57	1 14·3	1 14·5	1 10·9	5·7 0·4	11·7 0·9	17·7 1·3
58	1 14·5	1 14·7	1 11·1	5·8 0·4	11·8 0·9	17·8 1·3
59	1 14·8	1 15·0	1 11·3	5·9 0·4	11·9 0·9	17·9 1·3
60	1 15·0	1 15·2	1 11·6	6·0 0·5	12·0 0·9	18·0 1·4

5^m

5 (s)	SUN PLANETS	ARIES	MOON	v or d / Corrn	v or d / Corrn	v or d / Corrn
00	1 15·0	1 15·2	1 11·6	0·0 0·0	6·0 0·6	12·0 1·1
01	1 15·3	1 15·5	1 11·8	0·1 0·0	6·1 0·6	12·1 1·1
02	1 15·5	1 15·7	1 12·1	0·2 0·0	6·2 0·6	12·2 1·1
03	1 15·8	1 16·0	1 12·3	0·3 0·0	6·3 0·6	12·3 1·1
04	1 16·0	1 16·2	1 12·5	0·4 0·0	6·4 0·6	12·4 1·1
05	1 16·3	1 16·5	1 12·8	0·5 0·0	6·5 0·6	12·5 1·1
06	1 16·5	1 16·7	1 13·0	0·6 0·1	6·6 0·6	12·6 1·2
07	1 16·8	1 17·0	1 13·3	0·7 0·1	6·7 0·6	12·7 1·2
08	1 17·0	1 17·2	1 13·5	0·8 0·1	6·8 0·6	12·8 1·2
09	1 17·3	1 17·5	1 13·7	0·9 0·1	6·9 0·6	12·9 1·2
10	1 17·5	1 17·7	1 14·0	1·0 0·1	7·0 0·6	13·0 1·2
11	1 17·8	1 18·0	1 14·2	1·1 0·1	7·1 0·7	13·1 1·2
12	1 18·0	1 18·2	1 14·4	1·2 0·1	7·2 0·7	13·2 1·2
13	1 18·3	1 18·5	1 14·7	1·3 0·1	7·3 0·7	13·3 1·2
14	1 18·5	1 18·7	1 14·9	1·4 0·1	7·4 0·7	13·4 1·2
15	1 18·8	1 19·0	1 15·2	1·5 0·1	7·5 0·7	13·5 1·2
16	1 19·0	1 19·2	1 15·4	1·6 0·1	7·6 0·7	13·6 1·2
17	1 19·3	1 19·5	1 15·6	1·7 0·2	7·7 0·7	13·7 1·3
18	1 19·5	1 19·7	1 15·9	1·8 0·2	7·8 0·7	13·8 1·3
19	1 19·8	1 20·0	1 16·1	1·9 0·2	7·9 0·7	13·9 1·3
20	1 20·0	1 20·2	1 16·4	2·0 0·2	8·0 0·7	14·0 1·3
21	1 20·3	1 20·5	1 16·6	2·1 0·2	8·1 0·7	14·1 1·3
22	1 20·5	1 20·7	1 16·8	2·2 0·2	8·2 0·8	14·2 1·3
23	1 20·8	1 21·0	1 17·1	2·3 0·2	8·3 0·8	14·3 1·3
24	1 21·0	1 21·2	1 17·3	2·4 0·2	8·4 0·8	14·4 1·3
25	1 21·3	1 21·5	1 17·5	2·5 0·2	8·5 0·8	14·5 1·3
26	1 21·5	1 21·7	1 17·8	2·6 0·2	8·6 0·8	14·6 1·3
27	1 21·8	1 22·0	1 18·0	2·7 0·2	8·7 0·8	14·7 1·3
28	1 22·0	1 22·2	1 18·3	2·8 0·3	8·8 0·8	14·8 1·4
29	1 22·3	1 22·5	1 18·5	2·9 0·3	8·9 0·8	14·9 1·4
30	1 22·5	1 22·7	1 18·7	3·0 0·3	9·0 0·8	15·0 1·4
31	1 22·8	1 23·0	1 19·0	3·1 0·3	9·1 0·8	15·1 1·4
32	1 23·0	1 23·2	1 19·2	3·2 0·3	9·2 0·8	15·2 1·4
33	1 23·3	1 23·5	1 19·5	3·3 0·3	9·3 0·9	15·3 1·4
34	1 23·5	1 23·7	1 19·7	3·4 0·3	9·4 0·9	15·4 1·4
35	1 23·8	1 24·0	1 19·9	3·5 0·3	9·5 0·9	15·5 1·4
36	1 24·0	1 24·2	1 20·2	3·6 0·3	9·6 0·9	15·6 1·4
37	1 24·3	1 24·5	1 20·4	3·7 0·3	9·7 0·9	15·7 1·4
38	1 24·5	1 24·7	1 20·7	3·8 0·3	9·8 0·9	15·8 1·4
39	1 24·8	1 25·0	1 20·9	3·9 0·4	9·9 0·9	15·9 1·5
40	1 25·0	1 25·2	1 21·1	4·0 0·4	10·0 0·9	16·0 1·5
41	1 25·3	1 25·5	1 21·4	4·1 0·4	10·1 0·9	16·1 1·5
42	1 25·5	1 25·7	1 21·6	4·2 0·4	10·2 0·9	16·2 1·5
43	1 25·8	1 26·0	1 21·8	4·3 0·4	10·3 0·9	16·3 1·5
44	1 26·0	1 26·2	1 22·1	4·4 0·4	10·4 1·0	16·4 1·5
45	1 26·3	1 26·5	1 22·3	4·5 0·4	10·5 1·0	16·5 1·5
46	1 26·5	1 26·7	1 22·6	4·6 0·4	10·6 1·0	16·6 1·5
47	1 26·8	1 27·0	1 22·8	4·7 0·4	10·7 1·0	16·7 1·5
48	1 27·0	1 27·2	1 23·0	4·8 0·4	10·8 1·0	16·8 1·5
49	1 27·3	1 27·5	1 23·3	4·9 0·4	10·9 1·0	16·9 1·5
50	1 27·5	1 27·7	1 23·5	5·0 0·5	11·0 1·0	17·0 1·6
51	1 27·8	1 28·0	1 23·8	5·1 0·5	11·1 1·0	17·1 1·6
52	1 28·0	1 28·2	1 24·0	5·2 0·5	11·2 1·0	17·2 1·6
53	1 28·3	1 28·5	1 24·2	5·3 0·5	11·3 1·0	17·3 1·6
54	1 28·5	1 28·7	1 24·5	5·4 0·5	11·4 1·0	17·4 1·6
55	1 28·8	1 29·0	1 24·7	5·5 0·5	11·5 1·1	17·5 1·6
56	1 29·0	1 29·2	1 24·9	5·6 0·5	11·6 1·1	17·6 1·6
57	1 29·3	1 29·5	1 25·2	5·7 0·5	11·7 1·1	17·7 1·6
58	1 29·5	1 29·7	1 25·4	5·8 0·5	11·8 1·1	17·8 1·6
59	1 29·8	1 30·0	1 25·7	5·9 0·5	11·9 1·1	17·9 1·6
60	1 30·0	1 30·2	1 25·9	6·0 0·6	12·0 1·1	18·0 1·7

APPENDIX 2C

INCREMENTS AND CORRECTIONS

10	SUN PLANETS	ARIES	MOON	v or d	Corr^n	v or d	Corr^n	v or d	Corr^n	11	SUN PLANETS	ARIES	MOON	v or d	Corr^n	v or d	Corr^n	v or d	Corr^n
s	° ′	° ′	° ′	′	′	′	′	′	′	s	° ′	° ′	° ′	′	′	′	′	′	′
00	2 30·0	2 30·4	2 23·2	0·0	0·0	6·0	1·1	12·0	2·1	00	2 45·0	2 45·5	2 37·5	0·0	0·0	6·0	1·2	12·0	2·3
01	2 30·3	2 30·7	2 23·4	0·1	0·0	6·1	1·1	12·1	2·1	01	2 45·3	2 45·7	2 37·7	0·1	0·0	6·1	1·2	12·1	2·3
02	2 30·5	2 30·9	2 23·6	0·2	0·0	6·2	1·1	12·2	2·1	02	2 45·5	2 46·0	2 38·0	0·2	0·0	6·2	1·2	12·2	2·3
03	2 30·8	2 31·2	2 23·9	0·3	0·1	6·3	1·1	12·3	2·2	03	2 45·8	2 46·2	2 38·2	0·3	0·1	6·3	1·2	12·3	2·4
04	2 31·0	2 31·4	2 24·1	0·4	0·1	6·4	1·1	12·4	2·2	04	2 46·0	2 46·5	2 38·4	0·4	0·1	6·4	1·2	12·4	2·4
05	2 31·3	2 31·7	2 24·4	0·5	0·1	6·5	1·1	12·5	2·2	05	2 46·3	2 46·7	2 38·7	0·5	0·1	6·5	1·2	12·5	2·4
06	2 31·5	2 31·9	2 24·6	0·6	0·1	6·6	1·2	12·6	2·2	06	2 46·5	2 47·0	2 38·9	0·6	0·1	6·6	1·3	12·6	2·4
07	2 31·8	2 32·2	2 24·8	0·7	0·1	6·7	1·2	12·7	2·2	07	2 46·8	2 47·2	2 39·2	0·7	0·1	6·7	1·3	12·7	2·4
08	2 32·0	2 32·4	2 25·1	0·8	0·1	6·8	1·2	12·8	2·2	08	2 47·0	2 47·5	2 39·4	0·8	0·2	6·8	1·3	12·8	2·5
09	2 32·3	2 32·7	2 25·3	0·9	0·2	6·9	1·2	12·9	2·3	09	2 47·3	2 47·7	2 39·6	0·9	0·2	6·9	1·3	12·9	2·5
10	2 32·5	2 32·9	2 25·6	1·0	0·2	7·0	1·2	13·0	2·3	10	2 47·5	2 48·0	2 39·9	1·0	0·2	7·0	1·3	13·0	2·5
11	2 32·8	2 33·2	2 25·8	1·1	0·2	7·1	1·2	13·1	2·3	11	2 47·8	2 48·2	2 40·1	1·1	0·2	7·1	1·4	13·1	2·5
12	2 33·0	2 33·4	2 26·0	1·2	0·2	7·2	1·3	13·2	2·3	12	2 48·0	2 48·5	2 40·3	1·2	0·2	7·2	1·4	13·2	2·5
13	2 33·3	2 33·7	2 26·3	1·3	0·2	7·3	1·3	13·3	2·3	13	2 48·3	2 48·7	2 40·6	1·3	0·2	7·3	1·4	13·3	2·5
14	2 33·5	2 33·9	2 26·5	1·4	0·2	7·4	1·3	13·4	2·3	14	2 48·5	2 49·0	2 40·8	1·4	0·3	7·4	1·4	13·4	2·6
15	2 33·8	2 34·2	2 26·7	1·5	0·3	7·5	1·3	13·5	2·4	15	2 48·8	2 49·2	2 41·1	1·5	0·3	7·5	1·4	13·5	2·6
16	2 34·0	2 34·4	2 27·0	1·6	0·3	7·6	1·3	13·6	2·4	16	2 49·0	2 49·5	2 41·3	1·6	0·3	7·6	1·5	13·6	2·6
17	2 34·3	2 34·7	2 27·2	1·7	0·3	7·7	1·3	13·7	2·4	17	2 49·3	2 49·7	2 41·5	1·7	0·3	7·7	1·5	13·7	2·6
18	2 34·5	2 34·9	2 27·5	1·8	0·3	7·8	1·4	13·8	2·4	18	2 49·5	2 50·0	2 41·8	1·8	0·3	7·8	1·5	13·8	2·6
19	2 34·8	2 35·2	2 27·7	1·9	0·3	7·9	1·4	13·9	2·4	19	2 49·8	2 50·2	2 42·0	1·9	0·4	7·9	1·5	13·9	2·7
20	2 35·0	2 35·4	2 27·9	2·0	0·4	8·0	1·4	14·0	2·5	20	2 50·0	2 50·5	2 42·3	2·0	0·4	8·0	1·5	14·0	2·7
21	2 35·3	2 35·7	2 28·2	2·1	0·4	8·1	1·4	14 1	2·5	21	2 50·3	2 50·7	2 42·5	2·1	0·4	8·1	1·6	14·1	2·7
22	2 35·5	2 35·9	2 28·4	2·2	0·4	8·2	1·4	14·2	2·5	22	2 50·5	2 51·0	2 42·7	2·2	0·4	8·2	1·6	14·2	2·7
23	2 35·8	2 36·2	2 28·7	2·3	0·4	8·3	1·5	14·3	2·5	23	2 50·8	2 51·2	2 43·0	2·3	0·4	8·3	1·6	14·3	2·7
24	2 36·0	2 36·4	2 28·9	2·4	0·4	8·4	1·5	14·4	2·5	24	2 51·0	2 51·5	2 43·2	2·4	0·5	8·4	1·6	14·4	2·8
25	2 36·3	2 36·7	2 29·1	2·5	0·4	8·5	1·5	14·5	2·5	25	2 51·3	2 51·7	2 43·4	2·5	0·5	8·5	1·6	14·5	2·8
26	2 36·5	2 36·9	2 29·4	2·6	0·5	8·6	1·5	14·6	2·6	26	2 51·5	2 52·0	2 43·7	2·6	0·5	8·6	1·6	14·6	2·8
27	2 36·8	2 37·2	2 29·6	2·7	0·5	8·7	1·5	14·7	2·6	27	2 51·8	2 52·2	2 43·9	2·7	0·5	8·7	1·7	14·7	2·8
28	2 37·0	2 37·4	2 29·8	2·8	0·5	8·8	1·5	14·8	2·6	28	2 52·0	2 52·5	2 44·2	2·8	0·5	8·8	1·7	14·8	2·8
29	2 37·3	2 37·7	2 30·1	2·9	0·5	8·9	1·6	14·9	2·6	29	2 52·3	2 52·7	2 44·4	2·9	0·6	8·9	1·7	14·9	2·9
30	2 37·5	2 37·9	2 30·3	3·0	0·5	9·0	1·6	15·0	2·6	30	2 52·5	2 53·0	2 44·6	3·0	0·6	9·0	1·7	15·0	2·9
31	2 37·8	2 38·2	2 30·6	3·1	0·5	9·1	1·6	15·1	2·6	31	2 52·8	2 53·2	2 44·9	3·1	0·6	9·1	1·7	15·1	2·9
32	2 38·0	2 38·4	2 30·8	3·2	0·6	9·2	1·6	15·2	2·7	32	2 53·0	2 53·5	2 45·1	3·2	0·6	9·2	1·8	15·2	2·9
33	2 38·3	2 38·7	2 31·0	3·3	0·6	9·3	1·6	15·3	2·7	33	2 53·3	2 53·7	2 45·4	3·3	0·6	9·3	1·8	15·3	2·9
34	2 38·5	2 38·9	2 31·3	3·4	0·6	9·4	1·6	15·4	2·7	34	2 53·5	2 54·0	2 45·6	3·4	0·7	9·4	1·8	15·4	3·0
35	2 38·8	2 39·2	2 31·5	3·5	0·6	9·5	1·7	15·5	2·7	35	2 53·8	2 54·2	2 45·8	3·5	0·7	9·5	1·8	15·5	3·0
36	2 39·0	2 39·4	2 31·8	3·6	0·6	9·6	1·7	15·6	2·7	36	2 54·0	2 54·5	2 46·1	3·6	0·7	9·6	1·8	15·6	3·0
37	2 39·3	2 39·7	2 32·0	3·7	0·6	9·7	1·7	15·7	2·7	37	2 54·3	2 54·7	2 46·3	3·7	0·7	9·7	1·9	15·7	3·0
38	2 39·5	2 39·9	2 32·2	3·8	0·7	9·8	1·7	15·8	2·8	38	2 54·5	2 55·0	2 46·6	3·8	0·7	9·8	1·9	15·8	3·0
39	2 39·8	2 40·2	2 32·5	3·9	0·7	9·9	1·7	15·9	2·8	39	2 54·8	2 55·2	2 46·8	3·9	0·7	9·9	1·9	15·9	3·0
40	2 40·0	2 40·4	2 32·7	4·0	0·7	10·0	1·8	16·0	2·8	40	2 55·0	2 55·5	2 47·0	4·0	0·8	10·0	1·9	16·0	3·1
41	2 40·3	2 40·7	2 32·9	4·1	0·7	10·1	1·8	16·1	2·8	41	2 55·3	2 55·7	2 47·3	4·1	0·8	10·1	1·9	16·1	3·1
42	2 40·5	2 40·9	2 33·2	4·2	0·7	10·2	1·8	16·2	2·8	42	2 55·5	2 56·0	2 47·5	4·2	0·8	10·2	2·0	16·2	3·1
43	2 40·8	2 41·2	2 33·4	4·3	0·8	10·3	1·8	16·3	2·9	43	2 55·8	2 56·2	2 47·7	4·3	0·8	10·3	2·0	16·3	3·1
44	2 41·0	2 41·4	2 33·7	4·4	0·8	10·4	1·8	16·4	2·9	44	2 56·0	2 56·5	2 48·0	4·4	0·8	10·4	2·0	16·4	3·1
45	2 41·3	2 41·7	2 33·9	4·5	0·8	10·5	1·8	16·5	2·9	45	2 56·3	2 56·7	2 48·2	4·5	0·9	10·5	2·0	16·5	3·2
46	2 41·5	2 41·9	2 34·1	4·6	0·8	10·6	1·9	16·6	2·9	46	2 56·5	2 57·0	2 48·5	4·6	0·9	10·6	2·0	16·6	3·2
47	2 41·8	2 42·2	2 34·4	4·7	0·8	10·7	1·9	16·7	2·9	47	2 56·8	2 57·2	2 48·7	4·7	0·9	10·7	2·1	16·7	3·2
48	2 42·0	2 42·4	2 34·6	4·8	0·8	10·8	1·9	16·8	2·9	48	2 57·0	2 57·5	2 48·9	4·8	0·9	10·8	2·1	16·8	3·2
49	2 42·3	2 42·7	2 34·9	4·9	0·9	10·9	1·9	16·9	3·0	49	2 57·3	2 57·7	2 49·2	4·9	0·9	10·9	2·1	16·9	3·2
50	2 42·5	2 42·9	2 35·1	5·0	0·9	11·0	1·9	17·0	3·0	50	2 57·5	2 58·0	2 49·4	5·0	1·0	11·0	2·1	17·0	3·3
51	2 42·8	2 43·2	2 35·3	5·1	0·9	11·1	1·9	17·1	3·0	51	2 57·8	2 58·2	2 49·7	5·1	1·0	11·1	2·1	17·1	3·3
52	2 43·0	2 43·4	2 35·6	5·2	0·9	11·2	2·0	17·2	3·0	52	2 58·0	2 58·5	2 49·9	5·2	1·0	11·2	2·1	17·2	3·3
53	2 43·3	2 43·7	2 35·8	5·3	0·9	11·3	2·0	17·3	3·0	53	2 58·3	2 58·7	2 50·1	5·3	1·0	11·3	2·2	17·3	3·3
54	2 43·5	2 43·9	2 36·1	5·4	0·9	11·4	2·0	17·4	3·0	54	2 58·5	2 59·0	2 50·4	5·4	1·0	11·4	2·2	17·4	3·3
55	2 43·8	2 44·2	2 36·3	5·5	1·0	11·5	2·0	17·5	3·1	55	2 58·8	2 59·2	2 50·6	5·5	1·1	11·5	2·2	17·5	3·4
56	2 44·0	2 44·4	2 36·5	5·6	1·0	11·6	2·0	17·6	3·1	56	2 59·0	2 59·5	2 50·8	5·6	1·1	11·6	2·2	17·6	3·4
57	2 44·3	2 44·7	2 36·8	5·7	1·0	11·7	2·0	17·7	3·1	57	2 59·3	2 59·7	2 51·1	5·7	1·1	11·7	2·2	17·7	3·4
58	2 44·5	2 45·0	2 37·0	5·8	1·0	11·8	2·1	17·8	3·1	58	2 59·5	3 00·0	2 51·3	5·8	1·1	11·8	2·3	17·8	3·4
59	2 44·8	2 45·2	2 37·2	5·9	1·0	11·9	2·1	17·9	3·1	59	2 59·8	3 00·2	2 51·6	5·9	1·1	11·9	2·3	17·9	3·4
60	2 45·0	2 45·5	2 37·5	6·0	1·1	12·0	2·1	18·0	3·2	60	3 00·0	3 00·5	2 51·8	6·0	1·2	12·0	2·3	18·0	3·5

APPENDIX 2D

22ᵐ INCREMENTS AND CORRECTIONS 23ᵐ

22ᵐ	SUN PLANETS	ARIES	MOON	v or Corrⁿ d	v or Corrⁿ d	v or Corrⁿ d	23ᵐ	SUN PLANETS	ARIES	MOON	v or Corrⁿ d	v or Corrⁿ d	v or Corrⁿ d
s	° ′	° ′	° ′	′ ′	′ ′	′ ′	s	° ′	° ′	° ′	′ ′	′ ′	′ ′
00	5 30·0	5 30·9	5 15·0	0·0 0·0	6·0 2·3	12·0 4·5	00	5 45·0	5 45·9	5 29·3	0·0 0·0	6·0 2·4	12·0 4·7
01	5 30·3	5 31·2	5 15·2	0·1 0·0	6·1 2·3	12·1 4·5	01	5 45·3	5 46·2	5 29·5	0·1 0·0	6·1 2·4	12·1 4·7
02	5 30·5	5 31·4	5 15·4	0·2 0·1	6·2 2·3	12·2 4·6	02	5 45·5	5 46·4	5 29·8	0·2 0·1	6·2 2·4	12·2 4·8
03	5 30·8	5 31·7	5 15·7	0·3 0·1	6·3 2·4	12·3 4·6	03	5 45·8	5 46·7	5 30·0	0·3 0·1	6·3 2·5	12·3 4·8
04	5 31·0	5 31·9	5 15·9	0·4 0·2	6·4 2·4	12·4 4·7	04	5 46·0	5 46·9	5 30·2	0·4 0·2	6·4 2·5	12·4 4·9
05	5 31·3	5 32·2	5 16·2	0·5 0·2	6·5 2·4	12·5 4·7	05	5 46·3	5 47·2	5 30·5	0·5 0·2	6·5 2·5	12·5 4·9
06	5 31·5	5 32·4	5 16·4	0·6 0·2	6·6 2·5	12·6 4·7	06	5 46·5	5 47·4	5 30·7	0·6 0·2	6·6 2·6	12·6 4·9
07	5 31·8	5 32·7	5 16·6	0·7 0·3	6·7 2·5	12·7 4·8	07	5 46·8	5 47·7	5 31·0	0·7 0·3	6·7 2·6	12·7 5·0
08	5 32·0	5 32·9	5 16·9	0·8 0·3	6·8 2·6	12·8 4·8	08	5 47·0	5 48·0	5 31·2	0·8 0·3	6·8 2·7	12·8 5·0
09	5 32·3	5 33·2	5 17·1	0·9 0·3	6·9 2·6	12·9 4·8	09	5 47·3	5 48·2	5 31·4	0·9 0·4	6·9 2·7	12·9 5·1
10	5 32·5	5 33·4	5 17·4	1·0 0·4	7·0 2·6	13·0 4·9	10	5 47·5	5 48·5	5 31·7	1·0 0·4	7·0 2·7	13·0 5·1
11	5 32·8	5 33·7	5 17·6	1·1 0·4	7·1 2·7	13·1 4·9	11	5 47·8	5 48·7	5 31·9	1·1 0·4	7·1 2·8	13·1 5·1
12	5 33·0	5 33·9	5 17·8	1·2 0·5	7·2 2·7	13·2 5·0	12	5 48·0	5 49·0	5 32·1	1·2 0·5	7·2 2·8	13·2 5·2
13	5 33·3	5 34·2	5 18·1	1·3 0·5	7·3 2·7	13·3 5·0	13	5 48·3	5 49·2	5 32·4	1·3 0·5	7·3 2·9	13·3 5·2
14	5 33·5	5 34·4	5 18·3	1·4 0·5	7·4 2·8	13·4 5·0	14	5 48·5	5 49·5	5 32·6	1·4 0·5	7·4 2·9	13·4 5·2
15	5 33·8	5 34·7	5 18·5	1·5 0·6	7·5 2·8	13·5 5·1	15	5 48·8	5 49·7	5 32·9	1·5 0·6	7·5 2·9	13·5 5·3
16	5 34·0	5 34·9	5 18·8	1·6 0·6	7·6 2·9	13·6 5·1	16	5 49·0	5 50·0	5 33·1	1·6 0·6	7·6 3·0	13·6 5·3
17	5 34·3	5 35·2	5 19·0	1·7 0·6	7·7 2·9	13·7 5·1	17	5 49·3	5 50·2	5 33·3	1·7 0·7	7·7 3·0	13·7 5·4
18	5 34·5	5 35·4	5 19·3	1·8 0·7	7·8 2·9	13·8 5·2	18	5 49·5	5 50·5	5 33·6	1·8 0·7	7·8 3·1	13·8 5·4
19	5 34·8	5 35·7	5 19·5	1·9 0·7	7·9 3·0	13·9 5·2	19	5 49·8	5 50·7	5 33·8	1·9 0·7	7·9 3·1	13·9 5·4
20	5 35·0	5 35·9	5 19·7	2·0 0·8	8·0 3·0	14·0 5·3	20	5 50·0	5 51·0	5 34·1	2·0 0·8	8·0 3·1	14·0 5·5
21	5 35·3	5 36·2	5 20·0	2·1 0·8	8·1 3·0	14·1 5·3	21	5 50·3	5 51·2	5 34·3	2·1 0·8	8·1 3·2	14·1 5·5
22	5 35·5	5 36·4	5 20·2	2·2 0·8	8·2 3·1	14·2 5·3	22	5 50·5	5 51·5	5 34·5	2·2 0·9	8·2 3·2	14·2 5·6
23	5 35·8	5 36·7	5 20·5	2·3 0·9	8·3 3·1	14·3 5·4	23	5 50·8	5 51·7	5 34·8	2·3 0·9	8·3 3·3	14·3 5·6
24	5 36·0	5 36·9	5 20·7	2·4 0·9	8·4 3·2	14·4 5·4	24	5 51·0	5 52·0	5 35·0	2·4 0·9	8·4 3·3	14·4 5·6
25	5 36·3	5 37·2	5 20·9	2·5 0·9	8·5 3·2	14·5 5·4	25	5 51·3	5 52·2	5 35·2	2·5 1·0	8·5 3·3	14·5 5·7
26	5 36·5	5 37·4	5 21·2	2·6 1·0	8·6 3·2	14·6 5·5	26	5 51·5	5 52·5	5 35·5	2·6 1·0	8·6 3·4	14·6 5·7
27	5 36·8	5 37·7	5 21·4	2·7 1·0	8·7 3·3	14·7 5·5	27	5 51·8	5 52·7	5 35·7	2·7 1·1	8·7 3·4	14·7 5·8
28	5 37·0	5 37·9	5 21·6	2·8 1·1	8·8 3·3	14·8 5·6	28	5 52·0	5 53·0	5 36·0	2·8 1·1	8·8 3·4	14·8 5·8
29	5 37·3	5 38·2	5 21·9	2·9 1·1	8·9 3·3	14·9 5·6	29	5 52·3	5 53·2	5 36·2	2·9 1·1	8·9 3·5	14·9 5·8
30	5 37·5	5 38·4	5 22·1	3·0 1·1	9·0 3·4	15·0 5·6	30	5 52·5	5 53·5	5 36·4	3·0 1·2	9·0 3·5	15·0 5·9
31	5 37·8	5 38·7	5 22·4	3·1 1·2	9·1 3·4	15·1 5·7	31	5 52·8	5 53·7	5 36·7	3·1 1·2	9·1 3·6	15·1 5·9
32	5 38·0	5 38·9	5 22·6	3·2 1·2	9·2 3·5	15·2 5·7	32	5 53·0	5 54·0	5 36·9	3·2 1·3	9·2 3·6	15·2 6·0
33	5 38·3	5 39·2	5 22·8	3·3 1·2	9·3 3·5	15·3 5·7	33	5 53·3	5 54·2	5 37·2	3·3 1·3	9·3 3·6	15·3 6·0
34	5 38·5	5 39·4	5 23·1	3·4 1·3	9·4 3·5	15·4 5·8	34	5 53·5	5 54·5	5 37·4	3·4 1·3	9·4 3·7	15·4 6·0
35	5 38·8	5 39·7	5 23·3	3·5 1·3	9·5 3·6	15·5 5·8	35	5 53·8	5 54·7	5 37·6	3·5 1·4	9·5 3·7	15·5 6·1
36	5 39·0	5 39·9	5 23·6	3·6 1·4	9·6 3·6	15·6 5·9	36	5 54·0	5 55·0	5 37·9	3·6 1·4	9·6 3·8	15·6 6·1
37	5 39·3	5 40·2	5 23·8	3·7 1·4	9·7 3·6	15·7 5·9	37	5 54·3	5 55·2	5 38·1	3·7 1·4	9·7 3·8	15·7 6·1
38	5 39·5	5 40·4	5 24·0	3·8 1·4	9·8 3·7	15·8 5·9	38	5 54·5	5 55·5	5 38·4	3·8 1·5	9·8 3·8	15·8 6·2
39	5 39·8	5 40·7	5 24·3	3·9 1·5	9·9 3·7	15·9 6·0	39	5 54·8	5 55·7	5 38·6	3·9 1·5	9·9 3·9	15·9 6·2
40	5 40·0	5 40·9	5 24·5	4·0 1·5	10·0 3·8	16·0 6·0	40	5 55·0	5 56·0	5 38·8	4·0 1·6	10·0 3·9	16·0 6·3
41	5 40·3	5 41·2	5 24·7	4·1 1·5	10·1 3·8	16·1 6·0	41	5 55·3	5 56·2	5 39·1	4·1 1·6	10·1 4·0	16·1 6·3
42	5 40·5	5 41·4	5 25·0	4·2 1·6	10·2 3·8	16·2 6·1	42	5 55·5	5 56·5	5 39·3	4·2 1·6	10·2 4·0	16·2 6·3
43	5 40·8	5 41·7	5 25·2	4·3 1·6	10·3 3·9	16·3 6·1	43	5 55·8	5 56·7	5 39·5	4·3 1·7	10·3 4·0	16·3 6·4
44	5 41·0	5 41·9	5 25·5	4·4 1·7	10·4 3·9	16·4 6·2	44	5 56·0	5 57·0	5 39·8	4·4 1·7	10·4 4·1	16·4 6·4
45	5 41·3	5 42·2	5 25·7	4·5 1·7	10·5 3·9	16·5 6·2	45	5 56·3	5 57·2	5 40·0	4·5 1·8	10·5 4·1	16·5 6·5
46	5 41·5	5 42·4	5 25·9	4·6 1·7	10·6 4·0	16·6 6·2	46	5 56·5	5 57·5	5 40·3	4·6 1·8	10·6 4·2	16·6 6·5
47	5 41·8	5 42·7	5 26·2	4·7 1·8	10·7 4·0	16·7 6·3	47	5 56·8	5 57·7	5 40·5	4·7 1·8	10·7 4·2	16·7 6·5
48	5 42·0	5 42·9	5 26·4	4·8 1·8	10·8 4·1	16·8 6·3	48	5 57·0	5 58·0	5 40·7	4·8 1·9	10·8 4·2	16·8 6·6
49	5 42·3	5 43·2	5 26·7	4·9 1·8	10·9 4·1	16·9 6·3	49	5 57·3	5 58·2	5 41·0	4·9 1·9	10·9 4·3	16·9 6·6
50	5 42·5	5 43·4	5 26·9	5·0 1·9	11·0 4·1	17·0 6·4	50	5 57·5	5 58·5	5 41·2	5·0 2·0	11·0 4·3	17·0 6·7
51	5 42·8	5 43·7	5 27·1	5·1 1·9	11·1 4·2	17·1 6·4	51	5 57·8	5 58·7	5 41·5	5·1 2·0	11·1 4·3	17·1 6·7
52	5 43·0	5 43·9	5 27·4	5·2 2·0	11·2 4·2	17·2 6·5	52	5 58·0	5 59·0	5 41·7	5·2 2·0	11·2 4·4	17·2 6·7
53	5 43·3	5 44·2	5 27·6	5·3 2·0	11·3 4·2	17·3 6·5	53	5 58·3	5 59·2	5 41·9	5·3 2·1	11·3 4·4	17·3 6·8
54	5 43·5	5 44·4	5 27·9	5·4 2·0	11·4 4·3	17·4 6·5	54	5 58·5	5 59·5	5 42·2	5·4 2·1	11·4 4·5	17·4 6·8
55	5 43·8	5 44·7	5 28·1	5·5 2·1	11·5 4·3	17·5 6·6	55	5 58·8	5 59·7	5 42·4	5·5 2·2	11·5 4·5	17·5 6·9
56	5 44·0	5 44·9	5 28·3	5·6 2·1	11·6 4·4	17·6 6·6	56	5 59·0	6 00·0	5 42·6	5·6 2·2	11·6 4·5	17·6 6·9
57	5 44·3	5 45·2	5 28·6	5·7 2·1	11·7 4·4	17·7 6·6	57	5 59·3	6 00·2	5 42·9	5·7 2·2	11·7 4·6	17·7 6·9
58	5 44·5	5 45·4	5 28·8	5·8 2·2	11·8 4·4	17·8 6·7	58	5 59·5	6 00·5	5 43·1	5·8 2·3	11·8 4·6	17·8 7·0
59	5 44·8	5 45·7	5 29·0	5·9 2·2	11·9 4·5	17·9 6·7	59	5 59·8	6 00·7	5 43·4	5·9 2·3	11·9 4·7	17·9 7·0
60	5 45·0	5 45·9	5 29·3	6·0 2·3	12·0 4·5	18·0 6·8	60	6 00·0	6 01·0	5 43·6	6·0 2·4	12·0 4·7	18·0 7·1

32ᵐ	SUN PLANETS	ARIES	MOON	v or Corrⁿ d	v or Corrⁿ d	v or Corrⁿ d	33ᵐ	SUN PLANETS	ARIES	MOON	v or Corrⁿ d	v or Corrⁿ d	v or Corrⁿ d
s	° ′	° ′	° ′	′ ′	′ ′	′ ′	s	° ′	° ′	° ′	′ ′	′ ′	′ ′
00	8 00·0	8 01·3	7 38·1	0·0 0·0	6·0 3·3	12·0 6·5	00	8 15·0	8 16·4	7 52·5	0·0 0·0	6·0 3·4	12·0 6·7
01	8 00·3	8 01·6	7 38·4	0·1 0·1	6·1 3·3	12·1 6·6	01	8 15·3	8 16·6	7 52·7	0·1 0·1	6·1 3·4	12·1 6·8
02	8 00·5	8 01·8	7 38·6	0·2 0·1	6·2 3·4	12·2 6·6	02	8 15·5	8 16·9	7 52·9	0·2 0·1	6·2 3·5	12·2 6·8
03	8 00·8	8 02·1	7 38·8	0·3 0·2	6·3 3·4	12·3 6·7	03	8 15·8	8 17·1	7 53·2	0·3 0·2	6·3 3·5	12·3 6·9
04	8 01·0	8 02·3	7 39·1	0·4 0·2	6·4 3·5	12·4 6·7	04	8 16·0	8 17·4	7 53·4	0·4 0·2	6·4 3·6	12·4 6·9
05	8 01·3	8 02·6	7 39·3	0·5 0·3	6·5 3·5	12·5 6·8	05	8 16·3	8 17·6	7 53·6	0·5 0·3	6·5 3·6	12·5 7·0
06	8 01·5	8 02·8	7 39·6	0·6 0·3	6·6 3·6	12·6 6·8	06	8 16·5	8 17·9	7 53·9	0·6 0·3	6·6 3·7	12·6 7·0
07	8 01·8	8 03·1	7 39·8	0·7 0·4	6·7 3·6	12·7 6·9	07	8 16·8	8 18·1	7 54·1	0·7 0·4	6·7 3·7	12·7 7·1
08	8 02·0	8 03·3	7 40·0	0·8 0·4	6·8 3·7	12·8 6·9	08	8 17·0	8 18·4	7 54·4	0·8 0·4	6·8 3·8	12·8 7·1
09	8 02·3	8 03·6	7 40·3	0·9 0·5	6·9 3·7	12·9 7·0	09	8 17·3	8 18·6	7 54·6	0·9 0·5	6·9 3·9	12·9 7·2
10	8 02·5	8 03·8	7 40·5	1·0 0·5	7·0 3·8	13·0 7·0	10	8 17·5	8 18·9	7 54·8	1·0 0·6	7·0 3·9	13·0 7·3
11	8 02·8	8 04·1	7 40·8	1·1 0·6	7·1 3·8	13·1 7·1	11	8 17·8	8 19·1	7 55·1	1·1 0·6	7·1 4·0	13·1 7·3
12	8 03·0	8 04·3	7 41·0	1·2 0·7	7·2 3·9	13·2 7·2	12	8 18·0	8 19·4	7 55·3	1·2 0·7	7·2 4·0	13·2 7·4
13	8 03·3	8 04·6	7 41·2	1·3 0·7	7·3 4·0	13·3 7·2	13	8 18·3	8 19·6	7 55·6	1·3 0·7	7·3 4·1	13·3 7·4
14	8 03·5	8 04·8	7 41·5	1·4 0·8	7·4 4·0	13·4 7·3	14	8 18·5	8 19·9	7 55·8	1·4 0·8	7·4 4·1	13·4 7·5
15	8 03·8	8 05·1	7 41·7	1·5 0·8	7·5 4·1	13·5 7·3	15	8 18·8	8 20·1	7 56·0	1·5 0·8	7·5 4·2	13·5 7·5
16	8 04·0	8 05·3	7 42·0	1·6 0·9	7·6 4·1	13·6 7·4	16	8 19·0	8 20·4	7 56·3	1·6 0·9	7·6 4·2	13·6 7·6
17	8 04·3	8 05·6	7 42·2	1·7 0·9	7·7 4·2	13·7 7·4	17	8 19·3	8 20·6	7 56·5	1·7 0·9	7·7 4·3	13·7 7·6
18	8 04·5	8 05·8	7 42·4	1·8 1·0	7·8 4·2	13·8 7·5	18	8 19·5	8 20·9	7 56·7	1·8 1·0	7·8 4·4	13·8 7·7
19	8 04·8	8 06·1	7 42·7	1·9 1·0	7·9 4·3	13·9 7·5	19	8 19·8	8 21·1	7 57·0	1·9 1·1	7·9 4·4	13·9 7·8
20	8 05·0	8 06·3	7 42·9	2·0 1·1	8·0 4·3	14·0 7·6	20	8 20·0	8 21·4	7 57·2	2·0 1·1	8·0 4·5	14·0 7·8
21	8 05·3	8 06·6	7 43·1	2·1 1·1	8·1 4·4	14·1 7·6	21	8 20·3	8 21·6	7 57·5	2·1 1·2	8·1 4·5	14·1 7·9
22	8 05·5	8 06·8	7 43·4	2·2 1·2	8·2 4·4	14·2 7·7	22	8 20·5	8 21·9	7 57·7	2·2 1·2	8·2 4·6	14·2 7·9
23	8 05·8	8 07·1	7 43·6	2·3 1·2	8·3 4·5	14·3 7·7	23	8 20·8	8 22·1	7 57·9	2·3 1·3	8·3 4·6	14·3 8·0
24	8 06·0	8 07·3	7 43·9	2·4 1·3	8·4 4·6	14·4 7·8	24	8 21·0	8 22·4	7 58·2	2·4 1·3	8·4 4·7	14·4 8·0
25	8 06·3	8 07·6	7 44·1	2·5 1·4	8·5 4·6	14·5 7·9	25	8 21·3	8 22·6	7 58·4	2·5 1·4	8·5 4·7	14·5 8·1
26	8 06·5	8 07·8	7 44·3	2·6 1·4	8·6 4·7	14·6 7·9	26	8 21·5	8 22·9	7 58·7	2·6 1·5	8·6 4·8	14·6 8·2
27	8 06·8	8 08·1	7 44·6	2·7 1·5	8·7 4·7	14·7 8·0	27	8 21·8	8 23·1	7 58·9	2·7 1·5	8·7 4·9	14·7 8·2
28	8 07·0	8 08·3	7 44·8	2·8 1·5	8·8 4·8	14·8 8·0	28	8 22·0	8 23·4	7 59·1	2·8 1·6	8·8 4·9	14·8 8·3
29	8 07·3	8 08·6	7 45·1	2·9 1·6	8·9 4·8	14·9 8·1	29	8 22·3	8 23·6	7 59·4	2·9 1·6	8·9 5·0	14·9 8·3
30	8 07·5	8 08·8	7 45·3	3·0 1·6	9·0 4·9	15·0 8·1	30	8 22·5	8 23·9	7 59·6	3·0 1·7	9·0 5·0	15·0 8·4
31	8 07·8	8 09·1	7 45·5	3·1 1·7	9·1 4·9	15·1 8·2	31	8 22·8	8 24·1	7 59·8	3·1 1·7	9·1 5·1	15·1 8·4
32	8 08·0	8 09·3	7 45·8	3·2 1·7	9·2 5·0	15·2 8·2	32	8 23·0	8 24·4	8 00·1	3·2 1·8	9·2 5·1	15·2 8·5
33	8 08·3	8 09·6	7 46·0	3·3 1·8	9·3 5·0	15·3 8·3	33	8 23·3	8 24·6	8 00·3	3·3 1·8	9·3 5·2	15·3 8·5
34	8 08·5	8 09·8	7 46·2	3·4 1·8	9·4 5·1	15·4 8·3	34	8 23·5	8 24·9	8 00·6	3·4 1·9	9·4 5·2	15·4 8·6
35	8 08·8	8 10·1	7 46·5	3·5 1·9	9·5 5·1	15·5 8·4	35	8 23·8	8 25·1	8 00·8	3·5 2·0	9·5 5·3	15·5 8·7
36	8 09·0	8 10·3	7 46·7	3·6 2·0	9·6 5·2	15·6 8·5	36	8 24·0	8 25·4	8 01·0	3·6 2·0	9·6 5·4	15·6 8·7
37	8 09·3	8 10·6	7 47·0	3·7 2·0	9·7 5·3	15·7 8·5	37	8 24·3	8 25·6	8 01·3	3·7 2·1	9·7 5·4	15·7 8·8
38	8 09·5	8 10·8	7 47·2	3·8 2·1	9·8 5·3	15·8 8·6	38	8 24·5	8 25·9	8 01·5	3·8 2·1	9·8 5·5	15·8 8·8
39	8 09·8	8 11·1	7 47·4	3·9 2·1	9·9 5·4	15·9 8·6	39	8 24·8	8 26·1	8 01·8	3·9 2·2	9·9 5·5	15·9 8·9
40	8 10·0	8 11·3	7 47·7	4·0 2·2	10·0 5·4	16·0 8·7	40	8 25·0	8 26·4	8 02·0	4·0 2·2	10·0 5·6	16·0 8·9
41	8 10·3	8 11·6	7 47·9	4·1 2·2	10·1 5·5	16·1 8·7	41	8 25·3	8 26·6	8 02·2	4·1 2·3	10·1 5·6	16·1 9·0
42	8 10·5	8 11·8	7 48·2	4·2 2·3	10·2 5·5	16·2 8·8	42	8 25·5	8 26·9	8 02·5	4·2 2·3	10·2 5·7	16·2 9·0
43	8 10·8	8 12·1	7 48·4	4·3 2·3	10·3 5·6	16·3 8·8	43	8 25·8	8 27·1	8 02·7	4·3 2·4	10·3 5·8	16·3 9·1
44	8 11·0	8 12·3	7 48·6	4·4 2·4	10·4 5·6	16·4 8·9	44	8 26·0	8 27·4	8 02·9	4·4 2·5	10·4 5·8	16·4 9·2
45	8 11·3	8 12·6	7 48·9	4·5 2·4	10·5 5·7	16·5 8·9	45	8 26·3	8 27·6	8 03·2	4·5 2·5	10·5 5·9	16·5 9·2
46	8 11·5	8 12·8	7 49·1	4·6 2·5	10·6 5·7	16·6 9·0	46	8 26·5	8 27·9	8 03·4	4·6 2·6	10·6 5·9	16·6 9·3
47	8 11·8	8 13·1	7 49·3	4·7 2·5	10·7 5·8	16·7 9·0	47	8 26·8	8 28·1	8 03·7	4·7 2·6	10·7 6·0	16·7 9·3
48	8 12·0	8 13·3	7 49·6	4·8 2·6	10·8 5·9	16·8 9·1	48	8 27·0	8 28·4	8 03·9	4·8 2·7	10·8 6·0	16·8 9·4
49	8 12·3	8 13·6	7 49·8	4·9 2·7	10·9 5·9	16·9 9·2	49	8 27·3	8 28·6	8 04·1	4·9 2·7	10·9 6·1	16·9 9·4
50	8 12·5	8 13·8	7 50·1	5·0 2·7	11·0 6·0	17·0 9·2	50	8 27·5	8 28·9	8 04·4	5·0 2·8	11·0 6·1	17·0 9·5
51	8 12·8	8 14·1	7 50·3	5·1 2·8	11·1 6·0	17·1 9·3	51	8 27·8	8 29·1	8 04·6	5·1 2·8	11·1 6·2	17·1 9·5
52	8 13·0	8 14·3	7 50·5	5·2 2·8	11·2 6·1	17·2 9·3	52	8 28·0	8 29·4	8 04·9	5·2 2·9	11·2 6·3	17·2 9·6
53	8 13·3	8 14·6	7 50·8	5·3 2·9	11·3 6·1	17·3 9·4	53	8 28·3	8 29·6	8 05·1	5·3 3·0	11·3 6·3	17·3 9·7
54	8 13·5	8 14·9	7 51·0	5·4 2·9	11·4 6·2	17·4 9·4	54	8 28·5	8 29·9	8 05·3	5·4 3·0	11·4 6·4	17·4 9·7
55	8 13·8	8 15·1	7 51·3	5·5 3·0	11·5 6·2	17·5 9·5	55	8 28·8	8 30·1	8 05·6	5·5 3·1	11·5 6·4	17·5 9·8
56	8 14·0	8 15·4	7 51·5	5·6 3·0	11·6 6·3	17·6 9·5	56	8 29·0	8 30·4	8 05·8	5·6 3·1	11·6 6·5	17·6 9·8
57	8 14·3	8 15·6	7 51·7	5·7 3·1	11·7 6·3	17·7 9·6	57	8 29·3	8 30·6	8 06·1	5·7 3·2	11·7 6·5	17·7 9·9
58	8 14·5	8 15·9	7 52·0	5·8 3·1	11·8 6·4	17·8 9·6	58	8 29·5	8 30·9	8 06·3	5·8 3·2	11·8 6·6	17·8 9·9
59	8 14·8	8 16·1	7 52·2	5·9 3·2	11·9 6·4	17·9 9·7	59	8 29·8	8 31·1	8 06·5	5·9 3·3	11·9 6·6	17·9 10·0
60	8 15·0	8 16·4	7 52·5	6·0 3·3	12·0 6·5	18·0 9·8	60	8 30·0	8 31·4	8 06·8	6·0 3·4	12·0 6·7	18·0 10·1

58ᵐ   INCREMENTS AND CORRECTIONS   **59ᵐ**

58ᵐ s	SUN PLANETS	ARIES	MOON	v or Corrⁿ d	v or Corrⁿ d	v or Corrⁿ d
00	14 30·0	14 32·4	13 50·4	0·0 0·0	6·0 5·9	12·0 11·7
01	14 30·3	14 32·6	13 50·6	0·1 0·1	6·1 5·9	12·1 11·8
02	14 30·5	14 32·9	13 50·8	0·2 0·2	6·2 6·0	12·2 11·9
03	14 30·8	14 33·1	13 51·1	0·3 0·3	6·3 6·1	12·3 12·0
04	14 31·0	14 33·4	13 51·3	0·4 0·4	6·4 6·2	12·4 12·1
05	14 31·3	14 33·6	13 51·6	0·5 0·5	6·5 6·3	12·5 12·2
06	14 31·5	14 33·9	13 51·8	0·6 0·6	6·6 6·4	12·6 12·3
07	14 31·8	14 34·1	13 52·0	0·7 0·7	6·7 6·5	12·7 12·4
08	14 32·0	14 34·4	13 52·3	0·8 0·8	6·8 6·6	12·8 12·5
09	14 32·3	14 34·6	13 52·5	0·9 0·9	6·9 6·7	12·9 12·6
10	14 32·5	14 34·9	13 52·8	1·0 1·0	7·0 6·8	13·0 12·7
11	14 32·8	14 35·1	13 53·0	1·1 1·1	7·1 6·9	13·1 12·8
12	14 33·0	14 35·4	13 53·2	1·2 1·2	7·2 7·0	13·2 12·9
13	14 33·3	14 35·6	13 53·5	1·3 1·3	7·3 7·1	13·3 13·0
14	14 33·5	14 35·9	13 53·7	1·4 1·4	7·4 7·2	13·4 13·1
15	14 33·8	14 36·1	13 53·9	1·5 1·5	7·5 7·3	13·5 13·2
16	14 34·0	14 36·4	13 54·2	1·6 1·6	7·6 7·4	13·6 13·3
17	14 34·3	14 36·6	13 54·4	1·7 1·7	7·7 7·5	13·7 13·4
18	14 34·5	14 36·9	13 54·7	1·8 1·8	7·8 7·6	13·8 13·5
19	14 34·8	14 37·1	13 54·9	1·9 1·9	7·9 7·7	13·9 13·6
20	14 35·0	14 37·4	13 55·1	2·0 2·0	8·0 7·8	14·0 13·7
21	14 35·3	14 37·6	13 55·4	2·1 2·0	8·1 7·9	14·1 13·7
22	14 35·5	14 37·9	13 55·6	2·2 2·1	8·2 8·0	14·2 13·8
23	14 35·8	14 38·1	13 55·9	2·3 2·2	8·3 8·1	14·3 13·9
24	14 36·0	14 38·4	13 56·1	2·4 2·3	8·4 8·2	14·4 14·0
25	14 36·3	14 38·6	13 56·3	2·5 2·4	8·5 8·3	14·5 14·1
26	14 36·5	14 38·9	13 56·6	2·6 2·5	8·6 8·4	14·6 14·2
27	14 36·8	14 39·2	13 56·8	2·7 2·6	8·7 8·5	14·7 14·3
28	14 37·0	14 39·4	13 57·0	2·8 2·7	8·8 8·6	14·8 14·4
29	14 37·3	14 39·7	13 57·3	2·9 2·8	8·9 8·7	14·9 14·5
30	14 37·5	14 39·9	13 57·5	3·0 2·9	9·0 8·8	15·0 14·6
31	14 37·8	14 40·2	13 57·8	3·1 3·0	9·1 8·9	15·1 14·7
32	14 38·0	14 40·4	13 58·0	3·2 3·1	9·2 9·0	15·2 14·8
33	14 38·3	14 40·7	13 58·2	3·3 3·2	9·3 9·1	15·3 14·9
34	14 38·5	14 40·9	13 58·5	3·4 3·3	9·4 9·2	15·4 15·0
35	14 38·8	14 41·2	13 58·7	3·5 3·4	9·5 9·3	15·5 15·1
36	14 39·0	14 41·4	13 59·0	3·6 3·5	9·6 9·4	15·6 15·2
37	14 39·3	14 41·7	13 59·2	3·7 3·6	9·7 9·5	15·7 15·3
38	14 39·5	14 41·9	13 59·4	3·8 3·7	9·8 9·6	15·8 15·4
39	14 39·8	14 42·2	13 59·7	3·9 3·8	9·9 9·7	15·9 15·5
40	14 40·0	14 42·4	13 59·9	4·0 3·9	10·0 9·8	16·0 15·6
41	14 40·3	14 42·7	14 00·1	4·1 4·0	10·1 9·8	16·1 15·7
42	14 40·5	14 42·9	14 00·4	4·2 4·1	10·2 9·9	16·2 15·8
43	14 40·8	14 43·2	14 00·6	4·3 4·2	10·3 10·0	16·3 15·9
44	14 41·0	14 43·4	14 00·9	4·4 4·3	10·4 10·1	16·4 16·0
45	14 41·3	14 43·7	14 01·1	4·5 4·4	10·5 10·2	16·5 16·1
46	14 41·5	14 43·9	14 01·3	4·6 4·5	10·6 10·3	16·6 16·2
47	14 41·8	14 44·2	14 01·6	4·7 4·6	10·7 10·4	16·7 16·3
48	14 42·0	14 44·4	14 01·8	4·8 4·7	10·8 10·5	16·8 16·4
49	14 42·3	14 44·7	14 02·1	4·9 4·8	10·9 10·6	16·9 16·5
50	14 42·5	14 44·9	14 02·3	5·0 4·9	11·0 10·7	17·0 16·6
51	14 42·8	14 45·2	14 02·5	5·1 5·0	11·1 10·8	17·1 16·7
52	14 43·0	14 45·4	14 02·8	5·2 5·1	11·2 10·9	17·2 16·8
53	14 43·3	14 45·7	14 03·0	5·3 5·2	11·3 11·0	17·3 16·9
54	14 43·5	14 45·9	14 03·3	5·4 5·3	11·4 11·1	17·4 17·0
55	14 43·8	14 46·2	14 03·5	5·5 5·4	11·5 11·2	17·5 17·1
56	14 44·0	14 46·4	14 03·7	5·6 5·5	11·6 11·3	17·6 17·2
57	14 44·3	14 46·7	14 04·0	5·7 5·6	11·7 11·4	17·7 17·3
58	14 44·5	14 46·9	14 04·2	5·8 5·7	11·8 11·5	17·8 17·4
59	14 44·8	14 47·2	14 04·4	5·9 5·8	11·9 11·6	17·9 17·5
60	14 45·0	14 47·4	14 04·7	6·0 5·9	12·0 11·7	18·0 17·6

59ᵐ s	SUN PLANETS	ARIES	MOON	v or Corrⁿ d	v or Corrⁿ d	v or Corrⁿ d
00	14 45·0	14 47·4	14 04·7	0·0 0·0	6·0 6·0	12·0 11·9
01	14 45·3	14 47·7	14 04·9	0·1 0·1	6·1 6·0	12·1 12·0
02	14 45·5	14 47·9	14 05·2	0·2 0·2	6·2 6·1	12·2 12·1
03	14 45·8	14 48·2	14 05·4	0·3 0·3	6·3 6·2	12·3 12·2
04	14 46·0	14 48·4	14 05·6	0·4 0·4	6·4 6·3	12·4 12·3
05	14 46·3	14 48·7	14 05·9	0·5 0·5	6·5 6·4	12·5 12·4
06	14 46·5	14 48·9	14 06·1	0·6 0·6	6·6 6·5	12·6 12·5
07	14 46·8	14 49·2	14 06·4	0·7 0·7	6·7 6·6	12·7 12·6
08	14 47·0	14 49·4	14 06·6	0·8 0·8	6·8 6·7	12·8 12·7
09	14 47·3	14 49·7	14 06·8	0·9 0·9	6·9 6·8	12·9 12·8
10	14 47·5	14 49·9	14 07·1	1·0 1·0	7·0 6·9	13·0 12·9
11	14 47·8	14 50·2	14 07·3	1·1 1·1	7·1 7·0	13·1 13·0
12	14 48·0	14 50·4	14 07·5	1·2 1·2	7·2 7·1	13·2 13·1
13	14 48·3	14 50·7	14 07·8	1·3 1·3	7·3 7·2	13·3 13·2
14	14 48·5	14 50·9	14 08·0	1·4 1·4	7·4 7·3	13·4 13·3
15	14 48·8	14 51·2	14 08·3	1·5 1·5	7·5 7·4	13·5 13·4
16	14 49·0	14 51·4	14 08·5	1·6 1·6	7·6 7·5	13·6 13·5
17	14 49·3	14 51·7	14 08·7	1·7 1·7	7·7 7·6	13·7 13·6
18	14 49·5	14 51·9	14 09·0	1·8 1·8	7·8 7·7	13·8 13·7
19	14 49·8	14 52·2	14 09·2	1·9 1·9	7·9 7·8	13·9 13·8
20	14 50·0	14 52·4	14 09·5	2·0 2·0	8·0 7·9	14·0 13·9
21	14 50·3	14 52·7	14 09·7	2·1 2·1	8·1 8·0	14·1 14·0
22	14 50·5	14 52·9	14 09·9	2·2 2·2	8·2 8·1	14·2 14·1
23	14 50·8	14 53·2	14 10·2	2·3 2·3	8·3 8·2	14·3 14·2
24	14 51·0	14 53·4	14 10·4	2·4 2·4	8·4 8·3	14·4 14·3
25	14 51·3	14 53·7	14 10·6	2·5 2·5	8·5 8·4	14·5 14·4
26	14 51·5	14 53·9	14 10·9	2·6 2·6	8·6 8·5	14·6 14·5
27	14 51·8	14 54·2	14 11·1	2·7 2·7	8·7 8·6	14·7 14·6
28	14 52·0	14 54·4	14 11·4	2·8 2·8	8·8 8·7	14·8 14·7
29	14 52·3	14 54·7	14 11·6	2·9 2·9	8·9 8·8	14·9 14·8
30	14 52·5	14 54·9	14 11·8	3·0 3·0	9·0 8·9	15·0 14·9
31	14 52·8	14 55·2	14 12·1	3·1 3·1	9·1 9·0	15·1 15·0
32	14 53·0	14 55·4	14 12·3	3·2 3·2	9·2 9·1	15·2 15·1
33	14 53·3	14 55·7	14 12·6	3·3 3·3	9·3 9·2	15·3 15·2
34	14 53·5	14 55·9	14 12·8	3·4 3·4	9·4 9·3	15·4 15·3
35	14 53·8	14 56·2	14 13·0	3·5 3·5	9·5 9·4	15·5 15·4
36	14 54·0	14 56·4	14 13·3	3·6 3·6	9·6 9·5	15·6 15·5
37	14 54·3	14 56·7	14 13·5	3·7 3·7	9·7 9·6	15·7 15·6
38	14 54·5	14 56·9	14 13·8	3·8 3·8	9·8 9·7	15·8 15·7
39	14 54·8	14 57·2	14 14·0	3·9 3·9	9·9 9·8	15·9 15·8
40	14 55·0	14 57·5	14 14·2	4·0 4·0	10·0 9·9	16·0 15·9
41	14 55·3	14 57·7	14 14·5	4·1 4·1	10·1 10·0	16·1 16·0
42	14 55·5	14 58·0	14 14·7	4·2 4·2	10·2 10·1	16·2 16·1
43	14 55·8	14 58·2	14 14·9	4·3 4·3	10·3 10·2	16·3 16·2
44	14 56·0	14 58·5	14 15·2	4·4 4·4	10·4 10·3	16·4 16·3
45	14 56·3	14 58·7	14 15·4	4·5 4·5	10·5 10·4	16·5 16·4
46	14 56·5	14 59·0	14 15·7	4·6 4·6	10·6 10·5	16·6 16·5
47	14 56·8	14 59·2	14 15·9	4·7 4·7	10·7 10·6	16·7 16·6
48	14 57·0	14 59·5	14 16·1	4·8 4·8	10·8 10·7	16·8 16·7
49	14 57·3	14 59·7	14 16·4	4·9 4·9	10·9 10·8	16·9 16·8
50	14 57·5	15 00·0	14 16·6	5·0 5·0	11·0 10·9	17·0 16·9
51	14 57·8	15 00·2	14 16·9	5·1 5·1	11·1 11·0	17·1 17·0
52	14 58·0	15 00·5	14 17·1	5·2 5·2	11·2 11·1	17·2 17·1
53	14 58·3	15 00·7	14 17·3	5·3 5·3	11·3 11·2	17·3 17·2
54	14 58·5	15 01·0	14 17·6	5·4 5·4	11·4 11·3	17·4 17·3
55	14 58·8	15 01·2	14 17·8	5·5 5·5	11·5 11·4	17·5 17·4
56	14 59·0	15 01·5	14 18·0	5·6 5·6	11·6 11·5	17·6 17·5
57	14 59·3	15 01·7	14 18·3	5·7 5·7	11·7 11·6	17·7 17·6
58	14 59·5	15 02·0	14 18·5	5·8 5·8	11·8 11·7	17·8 17·7
59	14 59·8	15 02·2	14 18·8	5·9 5·9	11·9 11·8	17·9 17·8
60	15 00·0	15 02·5	14 19·0	6·0 6·0	12·0 11·9	18·0 17·9

APPENDIX 3A

1982 JANUARY 1, 2, 3 (FRI., SAT., SUN.)

G.M.T.	SUN G.H.A.	Dec.	MOON G.H.A.	v	Dec.	d	H.P.	Lat.	Twilight Naut.	Civil	Sunrise	Moonrise 1	2	3	4
d h	o '	o '	o '		o '		'	o	h m	h m		h m	h m	h m	h m
1 00	179 10.5	S23 02.7	114 06.3 12.5		S10 21.0 10.7		56.6	N 72	08 23	10 40	■	12 32	12 15	11 59	11 43
01	194 10.2	02.5	128 37.8 12.6		10 10.3 10.9		56.6	N 70	08 05	09 48	■	12 20	12 11	12 01	11 52
02	209 09.9	02.3	143 09.4 12.5		9 59.4 10.9		56.6	68	07 50	09 16	■	12 11	12 07	12 03	12 00
03	224 09.6 ··	02.1	157 40.9 12.5		9 48.5 10.9		56.7	66	07 37	08 53	10 27	12 03	12 04	12 05	12 06
04	239 09.3	01.9	172 12.4 12.6		9 37.6 11.0		56.7	64	07 26	08 34	09 49	11 56	12 01	12 06	12 11
05	254 09.0	01.7	186 44.0 12.6		9 26.6 11.1		56.7	62	07 17	08 18	09 22	11 50	11 58	12 07	12 16
06	269 08.7	S23 01.5	201 15.6 12.5		S 9 15.5 11.1		56.7	60	07 09	08 05	09 02	11 45	11 56	12 08	12 20
07	284 08.4	01.3	215 47.1 12.6		9 04.4 11.1		56.7	N 58	07 02	07 54	08 45	11 40	11 54	12 09	12 24
08	299 08.1	01.1	230 18.7 12.6		8 53.3 11.3		56.8	56	06 56	07 44	08 31	11 36	11 53	12 09	12 27
F 09	314 07.8 ··	00.9	244 50.3 12.6		8 42.0 11.2		56.8	54	06 50	07 36	08 19	11 32	11 51	12 10	12 30
R 10	329 07.5	00.7	259 21.9 12.6		8 30.8 11.4		56.8	52	06 44	07 28	08 08	11 29	11 50	12 11	12 33
I 11	344 07.2	00.5	273 53.5 12.5		8 19.4 11.5		56.9	50	06 39	07 20	07 59	11 26	11 49	12 11	12 35
D 12	359 06.9	S23 00.3	288 25.0 12.6		S 8 08.1 11.4		56.9	45	06 28	07 05	07 38	11 19	11 46	12 13	12 41
A 13	14 06.6	23 00.1	302 56.6 12.6		7 56.7 11.5		56.9	N 40	06 18	06 52	07 22	11 14	11 44	12 14	12 45
Y 14	29 06.3	22 59.9	317 28.2 12.6		7 45.2 11.5		57.0	35	06 09	06 40	07 08	11 09	11 42	12 15	12 49
15	44 06.0 ··	59.7	331 59.8 12.6		7 33.7 11.6		57.0	30	06 00	06 30	06 56	11 05	11 40	12 15	12 52
16	59 05.7	59.5	346 31.4 12.6		7 22.1 11.6		57.0	20	05 44	06 12	06 35	10 57	11 37	12 17	12 59
17	74 05.4	59.3	1 03.0 12.5		7 10.5 11.7		57.0	N 10	05 28	05 55	06 17	10 51	11 34	12 18	13 04
18	89 05.2	S22 59.1	15 34.5 12.6		S 6 58.8 11.7		57.1	0	05 12	05 38	06 00	10 44	11 32	12 19	13 09
19	104 04.9	58.9	30 06.1 12.6		6 47.1 11.7		57.1	S 10	04 53	05 20	05 43	10 38	11 29	12 21	13 14
20	119 04.6	58.7	44 37.7 12.6		6 35.4 11.8		57.1	20	04 31	05 00	05 25	10 32	11 26	12 22	13 20
21	134 04.3 ··	58.4	59 09.2 12.6		6 23.6 11.8		57.2	30	04 02	04 36	05 03	10 24	11 23	12 24	13 26
22	149 04.0	58.2	73 40.8 12.6		6 11.8 11.9		57.2	35	03 44	04 21	04 50	10 20	11 21	12 25	13 30
23	164 03.7	58.0	88 12.4 12.5		5 59.9 11.9		57.2	40	03 22	04 03	04 36	10 15	11 20	12 26	13 34
2 00	179 03.4	S22 57.8	102 43.9 12.5		S 5 48.0 11.9		57.2	45	02 52	03 41	04 18	10 09	11 17	12 27	13 39
01	194 03.1	57.6	117 15.4 12.5		5 36.1 12.0		57.3	S 50	02 08	03 12	03 56	10 02	11 14	12 28	13 45
02	209 02.8	57.4	131 47.0 12.5		5 24.1 12.0		57.3	52	01 42	02 57	03 45	09 59	11 13	12 29	13 48
03	224 02.5 ··	57.2	146 18.5 12.5		5 12.1 12.1		57.3	54	01 03	02 40	03 34	09 56	11 12	12 30	13 51
04	239 02.2	56.9	160 50.0 12.5		5 00.0 12.1		57.4	56	////	02 19	03 20	09 52	11 10	12 31	13 54
05	254 01.9	56.7	175 21.5 12.5		4 47.9 12.1		57.4	58	////	01 51	03 04	09 47	11 08	12 32	13 58
06	269 01.6	S22 56.5	189 53.0 12.4		S 4 35.8 12.2		57.4	S 60	////	01 08	02 44	09 43	11 07	12 33	14 02

G.M.T.	SUN G.H.A.	Dec.	MOON G.H.A.	v	Dec.	d	H.P.	Lat.	Sunset	Twilight Civil	Naut.	Moonset 1	2	3	4
07	284 01.3	56.3	204 24.4 12.5		4 23.6 12.2		57.5								
S 08	299 01.0	56.1	218 55.9 12.4		4 11.4 12.2		57.5	o	h m	h m	h m	h m	h m	h m	h m
A 09	314 00.8 ··	55.8	233 27.3 12.4		3 59.2 12.3		57.5	N 72	■	13 28	15 45	21 44	23 41	25 40	01 40
T 10	329 00.5	55.6	247 58.7 12.4		3 46.9 12.3		57.5	N 70	■	14 20	16 04	21 54	23 42	25 34	01 34
U 11	344 00.2	55.4	262 30.1 12.4		3 34.6 12.3		57.6	68	■	14 52	16 19	22 02	23 44	25 29	01 29
R 12	358 59.9	S22 55.2	277 01.5 12.4		S 3 22.3 12.4		57.6	66	13 42	15 16	16 31	22 08	23 45	25 24	01 24
D 13	13 59.6	54.9	291 32.9 12.3		3 09.9 12.4		57.6	64	14 19	15 34	16 42	22 13	23 46	25 20	01 20
A 14	28 59.3	54.7	306 04.2 12.4		2 57.5 12.4		57.7	62	14 46	15 50	16 51	22 18	23 46	25 17	01 17
Y 15	43 59.0 ··	54.5	320 35.6 12.3		2 45.1 12.4		57.7	60	15 06	16 03	16 59	22 22	23 47	25 14	01 14
16	58 58.7	54.3	335 06.9 12.3		2 32.7 12.5		57.7	N 58	15 23	16 14	17 06	22 26	23 48	25 12	01 12
17	73 58.4	54.0	349 38.2 12.2		2 20.2 12.5		57.8	56	15 37	16 24	17 13	22 29	23 48	25 10	01 10
18	88 58.1	S22 53.8	4 09.4 12.3		S 2 07.7 12.5		57.8	54	15 49	16 33	17 19	22 32	23 49	25 08	01 08
19	103 57.8	53.6	18 40.7 12.2		1 55.2 12.5		57.8	52	16 00	16 41	17 24	22 34	23 49	25 06	01 06
20	118 57.6	53.3	33 11.9 12.1		1 42.7 12.6		57.9	50	16 10	16 48	17 29	22 36	23 49	25 04	01 04
21	133 57.3 ··	53.1	47 43.0 12.2		1 30.1 12.6		57.9	45	16 30	17 03	17 40	22 41	23 50	25 01	01 01
22	148 57.0	52.9	62 14.2 12.1		1 17.5 12.6		57.9	N 40	16 46	17 16	17 50	22 46	23 51	24 58	00 58
23	163 56.7	52.6	76 45.3 12.1		1 04.9 12.6		57.9	35	17 00	17 28	17 59	22 49	23 51	24 55	00 55
3 00	178 56.4	S22 52.4	91 16.4 12.1		S 0 52.3 12.6		58.0	30	17 12	17 38	18 08	22 52	23 52	24 53	00 53
01	193 56.1	52.2	105 47.5 12.0		0 39.7 12.7		58.0	20	17 33	17 57	18 24	22 58	23 53	24 49	00 49
02	208 55.8	51.9	120 18.5 12.0		0 27.0 12.6		58.0	N 10	17 51	18 13	18 40	23 02	23 54	24 46	00 46
03	223 55.5 ··	51.7	134 49.5 12.0		0 14.4 12.7		58.1	0	18 08	18 30	18 56	23 07	23 54	24 43	00 43
04	238 55.2	51.5	149 20.5 12.0		S 0 01.7 12.7		58.1	S 10	18 25	18 48	19 15	23 11	23 55	24 40	00 40
05	253 54.9	51.2	163 51.5 11.9		N 0 11.0 12.7		58.1	20	18 43	19 08	19 37	23 16	23 55	24 36	00 36
06	268 54.7	S22 51.0	178 22.4 11.8		N 0 23.7 12.8		58.2	30	19 05	19 32	20 05	23 21	23 56	24 32	00 32
07	283 54.4	50.8	192 53.2 11.9		0 36.5 12.7		58.2	35	19 18	19 47	20 23	23 24	23 57	24 30	00 30
08	298 54.1	50.5	207 24.1 11.7		0 49.2 12.7		58.2	40	19 33	20 05	20 46	23 27	23 57	24 28	00 28
S 09	313 53.8 ··	50.3	221 54.8 11.8		1 01.9 12.8		58.3	45	19 50	20 27	21 16	23 31	23 58	24 25	00 25
U 10	328 53.5	50.0	236 25.6 11.7		1 14.7 12.7		58.3	S 50	20 12	20 55	21 59	23 36	23 58	24 21	00 21
N 11	343 53.2	49.8	250 56.3 11.7		1 27.4 12.8		58.3	52	20 22	21 10	22 24	23 38	23 58	24 20	00 20
D 12	358 52.9	S22 49.6	265 27.0 11.6		N 1 40.2 12.8		58.4	54	20 34	21 27	23 03	23 40	23 59	24 18	00 18
A 13	13 52.6	49.3	279 57.6 11.6		1 53.0 12.7		58.4	56	20 48	21 48	////	23 43	23 59	24 16	00 16
Y 14	28 52.4	49.1	294 28.2 11.6		2 05.7 12.8		58.4	58	21 04	22 15	////	23 45	23 59	24 14	00 14
15	43 52.1 ··	48.8	308 58.8 11.5		2 18.5 12.8		58.4	S 60	21 23	22 57	////	23 48	24 00	00 00	00 11
16	58 51.8	48.6	323 29.3 11.4		2 31.3 12.8		58.5								
17	73 51.5	48.3	337 59.7 11.4		2 44.1 12.8		58.5								
18	88 51.2	S22 48.1	352 30.1 11.4		N 2 56.9 12.7		58.5								
19	103 50.9	47.8	7 00.5 11.3		3 09.6 12.8		58.6								
20	118 50.6	47.6	21 30.8 11.2		3 22.4 12.8		58.6								
21	133 50.4 ··	47.3	36 01.0 11.3		3 35.2 12.8		58.6								
22	148 50.1	47.1	50 31.3 11.1		3 48.0 12.7		58.7								
23	163 49.8	46.8	65 01.4 11.1		4 00.7 12.8		58.7								

		SUN					MOON				
Day	Eqn. of Time 00h	12h	Mer. Pass.	Mer. Pass. Upper	Lower	Age	Phase				
	m s	m s	h m	h m	h m	d					
1	03 18	03 32	12 04	16 56	04 32	06					
2	03 46	04 00	12 04	17 43	05 19	07	◐				
3	04 14	04 28	12 04	18 31	06 07	08					

S.D. 16.3 d 0.2   S.D. 15.5   15.7   15.9

SUN and MOON

G.M.T.	SUN G.H.A.	SUN Dec.	MOON G.H.A.	v	MOON Dec.	d	H.P.
3 00	178 59.6	N23 00.3	37 15.8	13.2	S17 50.0	7.4	54.2
01	193 59.5	23 00.1	51 48.0	13.1	17 57.4	7.3	54.2
02	208 59.4	22 59.9	66 20.1	13.1	18 04.7	7.2	54.2
03	223 59.3	·· 59.7	80 52.2	13.1	18 11.9	7.1	54.2
04	238 59.2	59.5	95 24.3	13.1	18 19.0	7.0	54.2
05	253 59.0	59.3	109 56.4	13.0	18 26.0	7.0	54.2
06	268 58.9	N22 59.1	124 28.4	13.0	S18 33.0	6.8	54.1
07	283 58.8	58.9	139 00.4	12.9	18 39.8	6.8	54.1
S 08	298 58.7	58.7	153 32.3	12.9	18 46.6	6.7	54.1
A 09	313 58.6	·· 58.5	168 04.2	12.9	18 53.3	6.6	54.1
T 10	328 58.5	58.4	182 36.1	12.8	18 59.9	6.6	54.1
U 11	343 58.4	58.2	197 07.9	12.8	19 06.5	6.4	54.1
R 12	358 58.2	N22 58.0	211 39.7	12.8	S19 12.9	6.4	54.1
D 13	13 58.1	57.8	226 11.5	12.7	19 19.3	6.3	54.1
A 14	28 58.0	57.5	240 43.2	12.7	19 25.5	6.2	54.1
Y 15	43 57.9	·· 57.3	255 14.9	12.7	19 31.7	6.1	54.1
16	58 57.8	57.1	269 46.6	12.6	19 37.8	6.0	54.1
17	73 57.7	56.9	284 18.2	12.6	19 43.8	6.0	54.1
18	88 57.5	N22 56.7	298 49.8	12.6	S19 49.8	5.8	54.1
19	103 57.4	56.5	313 21.4	12.5	19 55.6	5.7	54.1
20	118 57.3	56.3	327 52.9	12.5	20 01.3	5.7	54.1
21	133 57.2	·· 56.1	342 24.4	12.5	20 07.0	5.6	54.1
22	148 57.1	55.9	356 55.9	12.4	20 12.6	5.4	54.0
23	163 57.0	55.7	11 27.3	12.4	20 18.0	5.4	54.0
4 00	178 56.9	N22 55.5	25 58.7	12.4	S20 23.4	5.3	54.0
01	193 56.8	55.3	40 30.1	12.3	20 28.7	5.2	54.0
02	208 56.6	55.1	55 01.4	12.3	20 33.9	5.1	54.0
03	223 56.5	·· 54.9	69 32.7	12.3	20 39.0	5.0	54.0
04	238 56.4	54.7	84 04.0	12.2	20 44.0	4.9	54.0
05	253 56.3	54.4	98 35.2	12.2	20 48.9	4.8	54.0
06	268 56.2	N22 54.2	113 06.4	12.2	S20 53.7	4.7	54.0
07	283 56.1	54.0	127 37.6	12.1	20 58.4	4.7	54.0
08	298 56.0	53.8	142 08.7	12.1	21 03.1	4.5	54.0
S 09	313 55.9	·· 53.6	156 39.8	12.1	21 07.6	4.4	54.0
U 10	328 55.7	53.4	171 10.9	12.0	21 12.0	4.4	54.0
N 11	343 55.6	53.2	185 41.9	12.1	21 16.4	4.2	54.0
D 12	358 55.5	N22 52.9	200 13.0	12.0	S21 20.6	4.1	54.0
A 13	13 55.4	52.7	214 44.0	11.9	21 24.7	4.1	54.0
Y 14	28 55.3	52.5	229 14.9	12.0	21 28.8	3.9	54.0
15	43 55.2	·· 52.3	243 45.9	11.9	21 32.7	3.9	54.0
16	58 55.1	52.1	258 16.8	11.8	21 36.6	3.7	54.0
17	73 55.0	51.8	272 47.6	11.9	21 40.3	3.7	54.0
18	88 54.9	N22 51.6	287 18.5	11.8	S21 44.0	3.5	54.0
19	103 54.7	51.4	301 49.3	11.8	21 47.5	3.5	54.0
20	118 54.6	51.2	316 20.1	11.8	21 51.0	3.3	54.0
21	133 54.5	·· 51.0	330 50.9	11.7	21 54.3	3.3	54.0
22	148 54.4	50.7	345 21.6	11.7	21 57.6	3.1	54.0
23	163 54.3	50.5	359 52.4	11.7	22 00.7	3.1	54.0
5 00	178 54.2	N22 50.3	14 23.1	11.6	S22 03.8	2.9	54.0
01	193 54.1	50.1	28 53.7	11.7	22 06.7	2.8	54.0
02	208 54.0	49.8	43 24.4	11.6	22 09.5	2.8	54.0
03	223 53.9	·· 49.6	57 55.0	11.6	22 12.3	2.6	54.0
04	238 53.8	49.4	72 25.6	11.6	22 14.9	2.5	54.0
05	253 53.6	49.2	86 56.2	11.6	22 17.4	2.4	54.0
06	268 53.5	N22 48.9	101 26.8	11.6	S22 19.8	2.4	54.0
07	283 53.4	48.7	115 57.4	11.5	22 22.2	2.2	54.0
08	298 53.3	48.5	130 27.9	11.5	22 24.4	2.1	54.0
M 09	313 53.2	·· 48.2	144 58.4	11.5	22 26.5	2.0	54.0
O 10	328 53.1	48.0	159 28.9	11.5	22 28.5	1.9	54.0
N 11	343 53.0	47.8	173 59.4	11.4	22 30.4	1.8	54.0
D 12	358 52.9	N22 47.5	188 29.8	11.5	S22 32.2	1.7	54.0
A 13	13 52.8	47.3	203 00.3	11.4	22 33.9	1.6	54.0
Y 14	28 52.7	47.1	217 30.7	11.4	22 35.5	1.5	54.0
15	43 52.6	·· 46.8	232 01.1	11.4	22 37.0	1.3	54.0
16	58 52.5	46.6	246 31.5	11.3	22 38.3	1.3	54.0
17	73 52.4	46.4	261 01.9	11.3	22 39.6	1.2	54.0
18	88 52.2	N22 46.1	275 32.2	11.4	S22 40.8	1.0	54.0
19	103 52.1	45.9	290 02.6	11.4	22 41.8	1.0	54.0
20	118 52.0	45.6	304 33.0	11.3	22 42.8	0.8	54.0
21	133 51.9	·· 45.4	319 03.3	11.3	22 43.6	0.8	54.0
22	148 51.8	45.2	333 33.6	11.3	22 44.4	0.6	54.0
23	163 51.7	44.9	348 03.9	11.3	22 45.0	0.5	54.0
	S.D. 15.8	d 0.2	S.D. 14.7		14.7		14.7

Twilight, Sunrise, Moonrise

Lat.	Naut.	Civil	Sunrise	3	4	5	6
N 72	□	□	□	■	■	■	■
N 70	□	□	□	■	■	■	■
68	□	□	□	20 31	■	■	■
66	////	////	00 39	19 49	21 18	22 31	23 11
64	////	////	01 46	19 21	20 39	21 44	22 28
62	////	////	02 20	19 00	20 12	21 14	21 59
60	////	01 10	02 45	18 42	19 51	20 51	21 37
N 58	////	01 52	03 04	18 28	19 34	20 32	21 19
56	////	02 20	03 21	18 15	19 20	20 17	21 04
54	01 04	02 41	03 34	18 05	19 07	20 03	20 51
52	01 43	02 58	03 46	17 55	18 56	19 52	20 40
50	02 09	03 13	03 57	17 47	18 47	19 42	20 30
45	02 53	03 42	04 19	17 29	18 26	19 20	20 08
N 40	03 22	04 04	04 36	17 14	18 10	19 03	19 51
35	03 45	04 21	04 51	17 01	17 56	18 48	19 37
30	04 03	04 36	05 03	16 51	17 44	18 35	19 24
20	04 31	05 00	05 25	16 32	17 23	18 14	19 03
N 10	04 54	05 20	05 43	16 16	17 05	17 55	18 44
0	05 12	05 38	06 01	16 01	16 49	17 37	18 27
S 10	05 29	05 55	06 18	15 46	16 32	17 20	18 09
20	05 44	06 12	06 36	15 31	16 15	17 01	17 51
30	06 00	06 30	06 56	15 13	15 54	16 40	17 29
35	06 09	06 40	07 08	15 02	15 42	16 27	17 17
40	06 18	06 52	07 22	14 50	15 29	16 13	17 02
45	06 28	07 05	07 39	14 36	15 13	15 56	16 45
S 50	06 39	07 21	07 59	14 19	14 53	15 35	16 24
52	06 44	07 28	08 08	14 11	14 44	15 25	16 13
54	06 50	07 36	08 19	14 02	14 33	15 13	16 02
56	06 56	07 44	08 31	13 52	14 22	15 00	15 49
58	07 02	07 54	08 45	13 40	14 08	14 45	15 34
S 60	07 09	08 05	09 02	13 27	13 52	14 27	15 16

Sunset, Twilight, Moonset

Lat.	Sunset	Civil	Naut.	3	4	5	6
N 72	□	□	□	■	■	■	■
N 70	□	□	□	■	■	■	■
68	□	□	□	23 46	■	■	■
66	23 24	////	////	00 25	00 28	00 38	01 07
64	22 21	////	////	00 45	00 57	01 18	01 54
62	21 47	////	////	01 02	01 19	01 45	02 25
60	21 23	22 56	////	01 15	01 36	02 06	02 48
N 58	21 03	22 15	////	01 27	01 51	02 23	03 06
56	20 47	21 48	////	01 38	02 04	02 38	03 22
54	20 34	21 27	23 02	01 47	02 15	02 51	03 35
52	20 22	21 10	22 24	01 55	02 25	03 02	03 46
50	20 12	20 55	21 59	02 02	02 34	03 12	03 57
45	19 50	20 27	21 15	02 18	02 52	03 32	04 18
N 40	19 32	20 05	20 46	02 31	03 08	03 49	04 36
35	19 18	19 47	20 24	02 42	03 21	04 03	04 50
30	19 05	19 32	20 05	02 52	03 32	04 16	05 03
20	18 44	19 08	19 37	03 09	03 51	04 37	05 25
N 10	18 25	18 48	19 15	03 24	04 08	04 55	05 44
0	18 08	18 30	18 57	03 37	04 24	05 12	06 01
S 10	17 51	18 14	18 40	03 51	04 40	05 29	06 19
20	17 33	17 57	18 24	04 06	04 57	05 48	06 38
30	17 12	17 39	18 08	04 23	05 17	06 09	06 59
35	17 00	17 28	18 00	04 33	05 28	06 21	07 12
40	16 46	17 17	17 51	04 44	05 41	06 36	07 26
45	16 30	17 04	17 41	04 57	05 57	06 53	07 44
S 50	16 10	16 48	17 29	05 14	06 16	07 13	08 05
52	16 01	16 41	17 24	05 22	06 25	07 23	08 15
54	15 50	16 33	17 19	05 32	06 35	07 35	08 27
56	15 38	16 24	17 13	05 40	06 47	07 48	08 40
58	15 24	16 15	17 07	05 51	07 00	08 02	08 55
S 60	15 07	16 03	17 00	06 03	07 16	08 20	09 13

SUN / MOON

Day	Eqn. of Time 00ʰ	12ʰ	Mer. Pass.	Mer. Pass. Upper	Lower	Age	Phase
3	04 01	04 07	12 04	22 13	09 49	12	○
4	04 12	04 18	12 04	23 01	10 36	13	
5	04 23	04 28	12 04	23 49	11 25	14	

ALTITUDE CORRECTION TABLES 10°-90°—SUN, STARS, PLANETS

SUN

OCT.—MAR. App. Alt.	Lower Limb	Upper Limb	APR.—SEPT. App. Alt.	Lower Limb	Upper Limb
9 34	+10.8	−21.5	9 39	+10.6	−21.2
9 45	+10.9	−21.4	9 51	+10.7	−21.1
9 56	+11.0	−21.3	10 03	+10.8	−21.0
10 08	+11.1	−21.2	10 15	+10.9	−20.9
10 21	+11.2	−21.1	10 27	+11.0	−20.8
10 34	+11.3	−21.0	10 40	+11.1	−20.7
10 47	+11.4	−20.9	10 54	+11.2	−20.6
11 01	+11.5	−20.8	11 08	+11.3	−20.5
11 15	+11.6	−20.7	11 23	+11.4	−20.4
11 30	+11.7	−20.6	11 38	+11.5	−20.3
11 46	+11.8	−20.5	11 54	+11.6	−20.2
12 02	+11.9	−20.4	12 10	+11.7	−20.1
12 19	+12.0	−20.3	12 28	+11.8	−20.0
12 37	+12.1	−20.2	12 46	+11.9	−19.9
12 55	+12.2	−20.1	13 05	+12.0	−19.8
13 14	+12.3	−20.0	13 24	+12.1	−19.7
13 35	+12.4	−19.9	13 45	+12.2	−19.6
13 56	+12.5	−19.8	14 07	+12.3	−19.5
14 18	+12.6	−19.7	14 30	+12.4	−19.4
14 42	+12.7	−19.6	14 54	+12.5	−19.3
15 06	+12.8	−19.5	15 19	+12.6	−19.2
15 32	+12.9	−19.4	15 46	+12.7	−19.1
15 59	+13.0	−19.3	16 14	+12.8	−19.0
16 28	+13.1	−19.2	16 44	+12.9	−18.9
16 59	+13.2	−19.1	17 15	+13.0	−18.8
17 32	+13.3	−19.0	17 48	+13.1	−18.7
18 06	+13.4	−18.9	18 24	+13.2	−18.6
18 42	+13.5	−18.8	19 01	+13.3	−18.5
19 21	+13.6	−18.7	19 42	+13.4	−18.4
20 03	+13.7	−18.6	20 25	+13.5	−18.3
20 48	+13.8	−18.5	21 11	+13.6	−18.2
21 35	+13.9	−18.4	22 00	+13.7	−18.1
22 26	+14.0	−18.3	22 54	+13.8	−18.0
23 22	+14.1	−18.2	23 51	+13.9	−17.9
24 21	+14.2	−18.1	24 53	+14.0	−17.8
25 26	+14.3	−18.0	26 00	+14.1	−17.7
26 36	+14.4	−17.9	27 13	+14.2	−17.6
27 52	+14.5	−17.8	28 33	+14.3	−17.5
29 15	+14.6	−17.7	30 00	+14.4	−17.4
30 46	+14.7	−17.6	31 35	+14.5	−17.3
32 26	+14.8	−17.5	33 20	+14.6	−17.2
34 17	+14.9	−17.4	35 17	+14.7	−17.1
36 20	+15.0	−17.3	37 26	+14.8	−17.0
38 36	+15.1	−17.2	39 50	+14.9	−16.9
41 08	+15.2	−17.1	42 31	+15.0	−16.8
43 59	+15.3	−17.0	45 31	+15.1	−16.7
47 10	+15.4	−16.9	48 55	+15.2	−16.6
50 46	+15.5	−16.8	52 44	+15.3	−16.5
54 49	+15.6	−16.7	57 02	+15.4	−16.4
59 23	+15.7	−16.6	61 51	+15.5	−16.3
64 30	+15.8	−16.5	67 17	+15.6	−16.2
70 12	+15.9	−16.4	73 16	+15.7	−16.1
76 26	+16.0	−16.3	79 43	+15.8	−16.0
83 05	+16.1	−16.2	86 32	+15.9	−15.9
90 00			90 00		

STARS AND PLANETS

App. Alt.	Corrⁿ	App. Alt.	Additional Corrⁿ
9 56	−5.3		**1982**
10 08	−5.2		**VENUS**
10 20	−5.1		Jan. 1–Jan. 3
10 33	−5.0		
10 46	−5.0	0	+0.5
11 00	−4.9	6	+0.6
11 14	−4.8	20	+0.7
11 29	−4.7	31	
11 45	−4.6		Jan. 4–Feb. 7
12 01	−4.5		
12 18	−4.4	0	+0.6
12 35	−4.3	4	+0.7
12 54	−4.2	12	+0.8
13 13	−4.1	22	
13 33	−4.0		Feb. 8–Feb. 14
13 54	−3.9		
14 16	−3.8	0	+0.5
14 40	−3.7	6	+0.6
15 04	−3.6	20	+0.7
15 30	−3.5	31	
15 57	−3.4		Feb. 15–Mar. 2
16 26	−3.3		
16 56	−3.2	0	+0.4
17 28	−3.1	11	+0.5
18 02	−3.0	41	
18 38	−2.9		Mar. 3–Mar. 28
19 17	−2.8		
19 58	−2.7	0	+0.3
20 42	−2.6	46	
21 28	−2.5		Mar. 29–May 12
22 19	−2.4		
23 13	−2.3	0	+0.2
24 11	−2.2	47	
25 14	−2.1		May 13–Dec. 31
26 22	−2.0		
27 36	−1.9	0	+0.1
28 56	−1.8	42	
30 24	−1.7		**MARS**
32 00	−1.6		Jan. 1–Jan. 30
33 45	−1.5		
35 40	−1.4	0	+0.1
37 48	−1.3	60	
40 08	−1.2		Jan. 31–June 18
42 44	−1.1		
45 36	−1.0	0	+0.2
48 47	−0.9	41	+0.1
52 18	−0.8	75	
56 11	−0.7		June 19–Dec. 31
60 28	−0.6		
65 08	−0.5	0	+0.1
70 11	−0.4	60	
75 34	−0.3		
81 13	−0.2		
87 03	−0.1		
90 00	0.0		

DIP

Ht. of Eye (m)	Corrⁿ	Ht. of Eye (ft.)	Ht. of Eye (m)	Corrⁿ
2.4	−2.8	8.0	1.0	−1.8
2.6	−2.9	8.6	1.5	−2.2
2.8	−3.0	9.2	2.0	−2.5
3.0	−3.1	9.8	2.5	−2.8
3.2	−3.2	10.5	3.0	−3.0
3.4	−3.3	11.2	See table ←	
3.6	−3.4	11.9		
3.8	−3.5	12.6	m	
4.0	−3.6	13.3	20	−7.9
4.3	−3.7	14.1	22	−8.3
4.5	−3.8	14.9	24	−8.6
4.7	−3.9	15.7	26	−9.0
5.0	−4.0	16.5	28	−9.3
5.2	−4.1	17.4		
5.5	−4.2	18.3	30	−9.6
5.8	−4.3	19.1	32	−10.0
6.1	−4.4	20.1	34	−10.3
6.3	−4.5	21.0	36	−10.6
6.6	−4.6	22.0	38	−10.8
6.9	−4.7	22.9		
7.2	−4.8	23.9	40	−11.1
7.5	−4.9	24.9	42	−11.4
7.9	−5.0	26.0	44	−11.7
8.2	−5.1	27.1	46	−11.9
8.5	−5.2	28.1	48	−12.2
8.8	−5.3	29.2	ft.	
9.2	−5.4	30.4	2	−1.4
9.5	−5.5	31.5	4	−1.9
9.9	−5.6	32.7	6	−2.4
10.3	−5.7	33.9	8	−2.7
10.6	−5.8	35.1	10	−3.1
11.0	−5.9	36.3	See table ←	
11.4	−6.0	37.6		
11.8	−6.1	38.9	ft.	
12.2	−6.2	40.1	70	−8.1
12.6	−6.3	41.5	75	−8.4
13.0	−6.4	42.8	80	−8.7
13.4	−6.5	44.2	85	−8.9
13.8	−6.6	45.5	90	−9.2
14.2	−6.7	46.9	95	−9.5
14.7	−6.8	48.4		
15.1	−6.9	49.8	100	−9.7
15.5	−7.0	51.3	105	−9.9
16.0	−7.1	52.8	110	−10.2
16.5	−7.2	54.3	115	−10.4
16.9	−7.3	55.8	120	−10.6
17.4	−7.4	57.4	125	−10.8
17.9	−7.5	58.9		
18.4	−7.6	60.5	130	−11.1
18.8	−7.7	62.1	135	−11.3
19.3	−7.8	63.8	140	−11.5
19.8	−7.9	65.4	145	−11.7
20.4	−8.0	67.1	150	−11.9
20.9	−8.1	68.8	155	−12.1
21.4		70.5		

App. Alt. = Apparent altitude = Sextant altitude corrected for index error and dip.
For daylight observations of Venus, see page 260.

N. Lat. { LHA greater than 180°....... Zn=Z
{ LHA less than 180°....... Zn=360—Z

DECLINATION (15°-29°) SA

LHA	15°			16°			17°			18°			19°			20°			21°			22°	
	Hc	d	Z	Hc	d	Z	Hc	d	Z	Hc	d	Z	Hc	d	Z	Hc	d	Z	Hc	d	Z	Hc	d
70	24 49	+33	90	25 22	+33	89	25 55	+33	88	26 28	+33	87	27 01	+32	86	27 33	+32	85	28 05	+31	84	28 36	+31
71	24 01	33	89	24 34	34	88	25 08	33	87	25 41	32	86	26 13	32	85	26 45	32	84	27 17	32	83	27 49	31
72	23 13	33	88	23 46	34	87	24 20	33	87	24 53	32	86	25 25	33	85	25 58	32	84	26 30	31	83	27 01	31
73	22 25	34	88	22 59	33	87	23 32	33	86	24 05	33	85	24 38	32	84	25 10	32	83	25 42	32	82	26 14	31
74	21 37	34	87	22 11	33	86	22 44	33	85	23 17	33	84	23 50	32	84	24 22	33	83	24 55	31	82	25 26	32
75	20 49	+34	87	21 23	+33	86	21 56	+34	85	22 30	+32	84	23 02	+33	83	23 35	+32	82	24 07	+32	81	24 39	+32
76	20 01	34	86	20 35	34	85	21 09	33	84	21 42	33	83	22 15	33	82	22 48	32	82	23 20	32	81	23 52	32
77	19 14	33	85	19 47	34	85	20 21	33	84	20 54	33	83	21 27	33	82	22 00	33	81	22 33	32	80	23 05	32
78	18 26	34	85	19 00	33	84	19 33	34	83	20 07	33	82	20 40	33	82	21 13	33	80	21 46	32	80	22 18	32
79	17 38	34	84	18 12	34	83	18 46	33	83	19 19	34	82	19 53	33	81	20 26	32	80	20 58	33	79	21 31	32
80	16 51	+34	84	17 25	+33	83	17 58	+34	82	18 32	+33	81	19 05	+34	80	19 39	+32	79	20 11	+33	78	20 44	+32
81	16 03	34	83	16 37	34	82	17 11	34	81	17 45	33	81	18 18	34	80	18 52	33	79	19 25	32	78	19 57	32
82	15 15	35	83	15 50	34	82	16 24	34	81	16 58	33	80	17 31	34	79	18 05	33	78	18 38	33	77	19 11	32
83	14 28	34	82	15 02	34	81	15 36	34	80	16 10	34	79	16 44	34	79	17 18	33	78	17 51	33	77	18 24	33
84	13 40	35	81	14 15	34	81	14 49	34	80	15 23	34	79	15 57	34	78	16 31	33	77	17 04	34	76	17 38	33
85	12 53	+35	81	13 28	+34	80	14 02	+34	79	14 36	+34	78	15 10	+34	77	15 44	+34	77	16 18	+33	76	16 51	+34
86	12 06	35	80	12 41	34	79	13 15	35	79	13 50	34	78	14 24	34	77	14 58	34	76	15 32	33	75	16 05	34
87	11 19	35	80	11 54	34	79	12 28	35	78	13 03	34	77	13 37	34	76	14 11	34	76	14 45	34	75	15 19	34
88	10 32	35	79	11 07	34	78	11 41	34	77	12 16	35	77	12 51	34	76	13 25	34	75	13 59	34	74	14 33	34
89	09 45	35	79	10 20	35	78	10 55	35	77	11 30	34	76	12 04	35	75	12 39	34	74	13 13	34	74	13 47	35
90	08 58	+35	78	09 33	+35	77	10 08	+35	76	10 43	+35	76	11 18	+35	75	11 53	+34	74	12 27	+35	73	13 02	+34
91	08 11	35	77	08 46	36	77	09 22	35	76	09 57	35	75	10 32	35	74	11 07	35	73	11 42	34	72	12 16	35
92	07 24	36	77	08 00	35	76	08 35	36	75	09 11	35	74	09 46	35	74	10 21	35	73	10 56	35	72	11 31	35
93	06 38	35	76	07 13	36	75	07 49	35	75	08 25	35	74	09 00	35	73	09 35	36	72	10 11	35	71	10 46	35
94	05 51	36	76	06 27	36	75	07 03	36	74	07 39	35	73	08 14	36	72	08 50	35	72	09 25	36	71	10 01	35
95	05 05	+36	75	05 41	+36	74	06 17	+36	73	06 53	+36	73	07 29	+35	72	08 04	+36	71	08 40	+36	70	09 16	+35
96	04 19	36	74	04 55	36	74	05 31	36	73	06 07	36	72	06 43	36	71	07 19	36	70	07 55	36	70	08 31	36
97	03 32	37	74	04 09	36	73	04 45	37	72	05 22	36	72	05 58	36	71	06 34	36	70	07 10	36	69	07 46	36
98	02 46	37	73	03 23	37	72	04 00	36	72	04 36	37	71	05 13	36	70	05 49	37	69	06 26	36	69	07 02	36
99	02 01	36	73	02 37	37	72	03 14	37	71	03 51	37	70	04 28	36	70	05 04	37	69	05 41	37	68	06 18	36
100	01 15	+37	72	01 52	+37	71	02 29	+37	71	03 06	+37	70	03 43	+37	69	04 20	+37	68	04 57	+37	67	05 34	+36
101	00 29	38	71	01 07	37	71	01 44	37	70	02 21	37	69	02 58	37	68	03 35	38	68	04 13	37	67	04 50	37
102	-0 16	38	71	00 22	37	70	00 59	37	69	01 36	38	69	02 14	37	68	02 51	38	67	03 29	37	66	04 06	37
103	-1 01	38	70	-0 23	37	69	00 14	38	69	00 52	38	68	01 30	37	67	02 07	38	66	02 45	38	66	03 23	37
104	-1 46	38	70	-1 08	38	69	-0 30	38	68	00 08	37	67	00 46	37	67	01 23	38	66	02 01	38	65	02 39	38
105	-2 31	+38	69	-1 53	+38	68	-1 15	+39	68	-0 36	+38	67	00 02	+38	66	00 40	+38	65	01 18	+38	64	01 56	+38
106	-3 16	38	68	-2 37	38	68	-1 59	39	67	-1 20	38	66	-0 42	38	65	-0 04	39	65	00 35	38	64	01 13	39
107	-4 00	39	68	-3 21	38	67	-2 43	39	66	-2 04	39	66	-1 25	38	65	-0 47	39	64	-0 08	39	63	00 31	38
108	-4 44	39	67	-4 05	38	66	-3 27	39	66	-2 48	39	65	-2 09	39	64	-1 30	39	63	-0 51	39	63	-0 12	39
109	-5 28	39	67	-4 49	39	66	-4 10	39	65	-3 31	39	64	-2 52	40	64	-2 12	39	63	-1 33	39	62	-0 54	40
110	-6 12	+39	66	-5 33	+40	65	-4 53	+39	64	-4 14	+40	64	-3 34	+39	63	-2 55	+40	62	-2 15	+39	61	-1 36	+40
111				-6 16	39	64	-5 37	40	64	-4 57	40	63	-4 17	40	62	-3 37	40	62	-2 57	40	61	-2 17	40
112							-6 19	40	63	-5 39	40	62	-4 59	40	62	-4 19	40	61	-3 39	40	60	-2 59	41
113										-6 22	41	62	-5 41	40	61	-5 01	41	60	-4 20	40	60	-3 40	41
114													-6 23	41	60	-5 42	41	60	-5 01	40	59	-4 21	41
115																-5 42	+41	58	-5 01	+41			
116																			-5 41	41			

LHA	15°			16°			17°			18°			19°			20°			21°			22°	
86	-5 51	36	104	-6 27	36	105	274																
85	-5 05	36	105	-5 41	36	106	-6 17	36	107	275													
84	-4 19	36	106	-4 55	36	106	-5 31	36	107	-6 07	36	108	276										
83	-3 32	37	106	-4 09	36	107	-4 45	37	108	-5 22	36	109	-5 58	36	109	277							
82	-2 46	37	107	-3 23	37	108	-4 00	36	108	-4 36	37	109	-5 13	36	110	-5 49	37	111	-6 26	36	112	278		
81	-2 01	36	107	-2 37	37	108	-3 14	37	109	-3 51	37	110	-4 28	36	110	-5 04	37	111	-5 41	37	112	-6 18	36
80	-1 15	37	108	-1 52	37	109	-2 29	37	109	-3 06	37	110	-3 43	37	111	-4 20	37	112	-4 57	37	113	-5 34	36
79	-0 29	38	109	-1 07	37	109	-1 44	37	110	-2 21	37	111	-2 58	37	112	-3 35	38	112	-4 13	37	113	-4 50	37
78	00 16	38	109	-0 22	37	110	-0 59	37	111	-1 36	38	111	-2 14	37	112	-2 51	38	113	-3 29	37	114	-4 06	37
77	01 01	38	110	00 23	37	111	-0 14	38	111	-0 52	38	112	-1 30	37	113	-2 07	38	114	-2 45	38	114	-3 23	37
76	01 46	38	110	01 08	38	111	00 30	38	112	-0 08	38	113	-0 46	37	113	-1 23	38	114	-2 01	38	115	-2 39	38
75	02 31	-38	111	01 53	-38	112	01 15	-39	112	00 36	-38	113	-0 02	-38	114	-0 40	-38	115	-1 18	-38	116	-1 56	-38
74	03 16	39	112	02 37	38	112	01 59	39	113	01 20	38	114	00 42	38	115	00 04	39	115	-0 35	38	116	-1 13	37
73	04 00	39	112	03 21	38	113	02 43	39	114	02 04	39	114	01 25	39	115	00 47	39	116	00 08	39	117	-0 31	38
72	04 44	39	113	04 05	38	114	03 27	39	114	02 48	39	115	02 09	39	116	01 30	39	117	00 51	39	117	00 12	39
71	05 28	39	113	04 49	39	114	04 10	39	115	03 31	39	116	02 52	40	116	02 12	39	117	01 33	39	118	00 54	40
70	06 12	-39	114	05 33	-40	115	04 53	-39	116	04 14	-40	116	03 34	-39	117	02 55	-40	118	02 15	-39	119	01 36	-40

| | 15° | | | 16° | | | 17° | | | 18° | | | 19° | | | 20° | | | 21° | | | 22° | |

S. Lat. { LHA greater than 180°........Zn=180—Z
{ LHA less than 180°...........Zn=180+Z

DECLINATION (15°-29°) CONT

226

LAT 37°

23°		24°			25°			26°			27°			28°			29°			LHA
d	Z	Hc	d	Z	Hc	d	Z	Hc	d	Z	Hc	d	Z	Hc	d	Z	Hc	d	Z	
'	°	° '	'	°	° '	'	°	° '	'	°	° '	'	°	° '	'	°	° '	'	°	°
+31	82	29 38	+30	81	30 08	+29	80	30 37	+29	79	31 06	+29	78	31 35	+28	77	32 03	+28	76	290
30	81	28 50	30	80	29 20	30	79	29 50	29	78	30 19	29	77	30 48	29	76	31 17	28	75	289
31	81	28 03	30	80	28 33	30	79	29 03	30	78	29 33	29	77	30 02	28	76	30 30	28	75	288
31	80	27 16	30	79	27 46	30	78	28 16	30	77	28 46	29	76	29 15	29	75	29 44	28	74	287
31	80	26 29	30	79	26 59	31	78	27 30	30	77	28 00	29	76	28 29	29	75	28 58	29	74	286
+31	79	25 42	+31	78	26 13	+30	77	26 43	+30	76	27 13	+30	75	27 43	+29	74	28 12	+29	74	285
31	79	24 55	31	78	25 26	31	77	25 57	30	76	26 27	30	75	26 57	29	74	27 26	29	73	284
31	78	24 08	31	77	24 39	31	76	25 10	31	75	25 41	30	74	26 11	29	74	26 40	30	73	283
32	78	23 22	31	77	23 53	31	76	24 24	31	75	24 55	30	74	25 25	30	73	25 55	29	72	282
32	77	22 35	31	76	23 06	32	75	23 38	31	74	24 09	30	73	24 39	30	73	25 09	30	72	281
+32	77	21 48	+32	76	22 20	+32	75	22 52	+31	74	23 23	+30	73	23 53	+31	72	24 24	+30	71	280
32	76	21 02	32	75	21 34	32	74	22 06	31	73	22 37	31	72	23 08	31	72	23 39	30	71	279
33	76	20 16	32	75	20 48	32	74	21 20	31	73	21 51	32	72	22 23	31	71	22 54	30	70	278
33	75	19 30	32	74	20 02	32	73	20 34	32	72	21 06	31	71	21 37	32	71	22 09	30	70	277
33	75	18 44	32	74	19 16	32	73	19 48	32	72	20 20	32	71	20 52	32	70	21 24	31	69	276
+33	74	17 58	+32	73	18 30	+33	72	19 03	+32	71	19 35	+32	70	20 07	+32	70	20 39	+31	69	275
33	73	17 12	33	73	17 45	33	72	18 18	32	71	18 50	32	70	19 22	32	69	19 54	32	68	274
33	73	16 26	34	72	17 00	33	72	17 33	32	71	18 05	33	69	18 38	32	69	19 10	32	68	273
33	72	15 41	33	72	16 14	33	71	16 47	33	70	17 20	33	69	17 53	33	68	18 26	32	67	272
33	72	14 55	34	71	15 29	34	70	16 03	33	69	16 36	33	68	17 09	33	68	17 42	32	67	271
+34	71	14 10	+34	70	14 44	+34	70	15 18	+33	69	15 51	+34	68	16 25	+33	67	16 58	+33	66	270
34	71	13 25	34	70	13 59	34	69	14 33	34	68	15 07	34	67	15 41	33	67	16 14	33	66	269
34	70	12 40	35	69	13 15	34	69	13 49	34	68	14 23	34	67	14 57	34	66	15 31	33	65	268
34	70	11 55	35	69	12 30	35	68	13 05	34	67	13 39	34	66	14 13	34	66	14 47	34	65	267
35	69	11 11	35	68	11 46	35	67	12 21	34	67	12 55	35	66	13 30	34	65	14 04	34	64	266
+35	69	10 26	+36	68	11 02	+35	67	11 37	+35	66	12 12	+34	65	12 46	+35	64	13 21	+34	64	265
35	68	09 42	36	67	10 18	35	66	10 53	35	66	11 28	35	65	12 03	35	64	12 38	35	63	264
36	67	08 58	36	67	09 34	35	66	10 09	36	65	10 45	35	64	11 20	36	63	11 56	35	63	263
36	67	08 14	36	66	08 50	36	65	09 26	36	65	10 02	36	64	10 38	35	63	11 13	36	62	262
36	66	07 30	37	66	08 07	36	65	08 43	36	64	09 19	36	63	09 55	36	62	10 31	36	62	261
+37	66	06 47	+37	65	07 24	+36	64	08 00	+36	63	08 36	+37	63	09 13	+36	62	09 49	+36	61	260
37	65	06 04	36	64	06 41	36	64	07 17	37	63	07 54	37	62	08 31	36	61	09 07	37	60	259
38	65	05 21	37	64	05 58	37	63	06 35	37	62	07 12	37	62	07 49	37	60	08 26	36	60	258
38	64	04 38	37	63	05 15	38	63	05 53	37	62	06 30	37	61	07 07	37	60	07 44	37	59	257
38	63	03 55	38	63	04 33	37	62	05 10	38	61	05 48	38	60	06 26	37	60	07 03	38	59	256
+38	63	03 12	+39	62	03 51	+38	61	04 29	+38	61	05 07	+37	60	05 44	+38	59	06 22	+38	58	255
38	62	02 29	38	62	03 09	38	61	03 47	38	60	04 25	39	59	05 04	38	58	05 42	38	58	254
39	62	01 48	39	61	02 27	39	60	03 06	38	59	03 44	39	59	04 23	38	58	05 01	39	57	253
39	61	01 06	39	60	01 45	39	60	02 24	39	59	03 03	39	58	03 42	39	57	04 21	39	57	252
39	61	00 25	39	60	01 04	40	59	01 44	39	58	02 23	39	57	03 02	39	57	03 41	39	56	251
+40	60	−0 16	+39	59	00 23	+40	58	01 03	+40	58	01 43	+39	57	02 22	+40	56	03 02	+39	55	250
40	59	−0 57	40	59	−0 17	40	57	00 23	40	57	01 03	40	56	01 43	40	56	02 23	39	55	249
40	59	−1 38	40	58	−0 58	41	57	−0 17	40	56	00 23	40	56	01 03	40	55	01 43	41	54	248
40	58	−2 19	41	57	−1 38	41	57	−0 57	40	56	−0 17	41	55	00 24	41	54	01 05	40	54	247
41	57	−2 59	41	57	−2 18	41	56	−1 37	41	55	−0 56	41	54	−0 15	41	54	00 26	41	53	246
+41	57	−3 39	+42	56	−2 57	+41	55	−2 16	+41	55	−1 35	+42	54	−0 53	+41	53	−0 12	+42	52	245
42	56	−4 18	41	55	−3 37	42	55	−2 55	42	54	−2 13	42	53	−1 31	41	53	−0 50	42	52	244
42	56	−4 58	42	55	−4 16	43	54	−3 33	42	53	−2 51	42	53	−2 09	42	52	−1 27	42	51	243
118		−5 37	43	54	−4 54	42	53	−4 12	43	53	−3 29	42	52	−2 47	43	51	−2 04	42	51	242
119		−6 15	42	54	−5 33	43	53	−4 50	43	52	−4 07	43	51	−3 24	43	51	−2 41	43	50	241
120					−6 11	+44	52	−5 27	+43	51	−4 44	+43	51	−4 01	+43	50	−3 18	+44	49	240
121								−6 05	44	51	−5 21	43	50	−4 38	44	49	−3 54	44	49	239
122											−5 58	44	49	−5 14	44	49	−4 30	44	48	238
123														−5 50	45	48	−5 05	44	47	237
124																	−5 40	44	47	236

23°		24°			25°			26°			27°			28°			29°			
−37	114	280																		
37	115	−6 04	37	116	281															
38	115	−5 21	37	116	−5 58	37	117	282												
38	116	−4 38	37	117	−5 15	38	118	−5 53	37	118	283									
38	117	−3 55	38	117	−4 33	37	118	−5 10	38	119	−5 48	38	120	284						
−38	117	−3 12	−39	118	−3 51	−38	119	−4 29	−38	119	−5 07	−37	120	−5 44	−38	121	285			
38	118	−2 30	39	118	−3 09	38	119	−3 47	38	120	−4 25	39	121	−5 04	38	122	−5 42	38	122	286
39	118	−1 48	39	119	−2 27	39	120	−3 06	38	121	−3 44	39	121	−4 23	38	122	−5 01	39	123	287
39	119	−1 06	39	120	−1 45	39	120	−2 24	39	121	−3 03	39	122	−3 42	39	123	−4 21	39	123	288
39	119	−0 25	39	120	−1 04	40	121	−1 44	39	122	−2 23	39	123	−3 02	39	124	−3 41	39	124	289
−40	120	00 16	−39	121	−0 23	−40	122	−1 03	−40	122	−1 43	−39	123	−2 22	−40	124	−3 02	−39	125	290

23°	24°	25°	26°	27°	28°	29°

NAME TO LATITUDE

N. Lat. {LHA greater than 180°........ Zn=Z / LHA less than 180°............ Zn=360−Z

DECLINATION (15°-29°) SA

LHA	15° Hc	d	Z	16° Hc	d	Z	17° Hc	d	Z	18° Hc	d	Z	19° Hc	d	Z	20° Hc	d	Z	21° Hc	d	Z	22° Hc	d
0	88 00	+60	180	89 00	+60	180	90 00	-60	..	89 00	-60	0	88 00	-60	0	87 00	-60	0	86 00	-60	0	85 00	-60
1	87 47	50	154	88 37	26	136	89 03	26	90	88 37	50	44	87 47	56	25	86 51	58	17	85 53	58	13	84 55	59
2	87 14	36	136	87 50	15	117	88 05	14	90	87 51	37	62	87 14	47	43	86 27	52	32	85 35	56	25	84 39	56
3	86 30	27	124	86 57	11	109	87 08	10	90	86 58	27	70	86 31	39	55	85 52	46	43	85 06	51	35	84 15	53
4	85 40	22	117	86 02	9	104	86 11	8	89	86 03	21	75	85 42	32	62	85 10	40	51	84 30	46	43	83 44	49
5	84 48	+18	112	85 06	+7	101	85 13	-5	89	85 08	-17	77	84 51	-28	66	84 23	-34	57	83 49	-41	49	83 08	-45
6	83 54	16	108	84 10	6	99	84 16	4	89	84 12	15	79	83 57	23	70	83 34	30	61	83 04	37	54	82 27	41
7	82 59	14	106	83 13	5	98	83 18	3	89	83 15	12	80	83 03	20	72	82 43	27	65	82 16	33	58	81 43	37
8	82 03	13	104	82 16	5	96	82 21	3	89	82 18	10	81	82 08	17	74	81 51	24	67	81 27	30	61	80 57	34
9	81 07	12	102	81 19	5	95	81 24	2	89	81 22	9	82	81 13	16	76	80 57	21	69	80 36	27	63	80 09	31
10	80 11	+11	100	80 22	+4	95	80 26	-1	89	80 25	-8	83	80 17	-14	77	80 03	-19	71	79 44	-24	66	79 20	-29
11	79 14	10	99	79 24	5	94	79 29	1	88	79 28	7	83	79 21	12	78	79 09	17	72	78 52	22	67	78 30	27
12	78 18	9	98	78 27	5	93	78 32	-1	88	78 31	6	83	78 25	11	78	78 14	15	73	77 59	21	69	77 38	24
13	77 21	9	97	77 30	4	93	77 34	0	88	77 34	5	83	77 29	10	79	77 19	14	74	77 05	19	70	76 46	22
14	76 24	9	97	76 33	4	92	76 37	0	88	76 37	5	84	76 32	8	79	76 24	13	75	76 11	17	71	75 54	21
15	75 27	+8	96	75 35	+5	92	75 40	0	88	75 40	-4	84	75 36	-8	80	75 28	-12	76	75 16	-15	72	75 01	-19
16	74 30	8	95	74 38	4	91	74 42	+1	88	74 43	3	84	74 40	8	80	74 32	10	76	74 22	15	73	74 07	17
17	73 33	8	95	73 41	4	91	73 45	1	88	73 46	3	84	73 43	6	80	73 37	10	77	73 27	13	73	73 14	17
18	72 35	8	94	72 43	5	91	72 48	1	87	72 49	3	84	72 46	5	81	72 41	9	77	72 32	12	74	72 20	16
19	71 38	8	94	71 46	4	90	71 50	2	87	71 52	2	84	71 50	5	81	71 45	8	78	71 37	12	74	71 25	14
20	70 41	+7	93	70 48	+5	90	70 53	+2	87	70 55	-2	84	70 53	-4	81	70 49	-8	78	70 41	-10	75	70 31	-13
21	69 44	7	93	69 51	5	90	69 56	2	87	69 58	-2	84	69 56	3	81	69 53	7	78	69 46	10	75	69 36	12
22	68 46	8	92	68 54	4	90	68 58	2	87	69 00	0	84	69 00	4	81	68 56	6	78	68 50	9	76	68 41	11
23	67 49	7	92	67 56	5	89	68 01	4	87	68 03	0	84	68 03	3	81	68 00	5	79	67 55	8	76	67 47	11
24	66 52	7	92	66 59	5	89	67 04	2	86	67 06	0	84	67 06	2	81	67 04	5	79	66 59	7	76	66 52	10
25	65 54	+8	91	66 02	+5	89	66 07	+2	86	66 09	+1	84	66 10	-2	81	66 08	-5	79	66 03	-7	76	65 56	-9
26	64 57	7	91	65 04	5	89	65 09	3	86	65 12	1	84	65 13	2	81	65 11	4	79	65 07	6	77	65 01	8
27	63 59	8	91	64 07	5	88	64 12	3	86	64 15	1	84	64 16	1	81	64 15	3	79	64 12	6	77	64 06	8
28	63 02	7	90	63 09	6	88	63 15	3	86	63 18	1	84	63 19	0	81	63 19	3	79	63 16	5	77	63 11	7
29	62 05	7	90	62 12	6	88	62 18	3	86	62 21	2	84	62 23	-1	81	62 22	2	79	62 20	5	77	62 15	6
30	61 07	+8	90	61 15	+5	88	61 20	+4	86	61 24	+2	83	61 26	0	81	61 26	-2	79	61 24	-4	77	61 20	-6
31	60 10	7	89	60 17	6	87	60 23	4	85	60 27	2	83	60 29	1	81	60 29	1	79	60 28	4	77	60 24	5
32	59 13	7	89	59 20	6	87	59 26	4	85	59 30	3	83	59 33	0	81	59 33	1	79	59 32	3	77	59 29	5
33	58 15	8	89	58 23	6	87	58 29	4	85	58 33	3	83	58 36	+1	81	58 37	-1	79	58 36	3	77	58 33	4
34	57 18	8	89	57 26	6	87	57 32	4	85	57 36	3	83	57 39	1	81	57 40	0	79	57 40	2	77	57 38	4
35	56 20	+8	88	56 28	+7	87	56 35	+4	85	56 39	+3	83	56 42	+2	81	56 44	0	79	56 44	-2	78	56 42	-3
36	55 23	8	88	55 31	6	86	55 37	5	85	55 42	4	83	55 46	2	81	55 48	0	79	55 48	1	78	55 47	3
37	54 26	8	88	54 34	6	86	54 40	6	84	54 45	4	83	54 49	2	81	54 51	+1	79	54 52	1	78	54 51	2
38	53 28	9	88	53 37	6	86	53 43	6	84	53 49	3	83	53 52	3	81	53 55	1	79	53 56	-1	78	53 55	1
39	52 31	8	88	52 39	7	86	52 46	6	84	52 52	4	82	52 56	3	81	52 59	1	79	53 00	0	78	53 00	2
40	51 34	+8	87	51 42	+7	86	51 49	+6	84	51 55	+4	82	51 59	+3	81	52 02	+2	79	52 04	0	77	52 04	-1
41	50 36	9	87	50 45	7	85	50 52	6	84	50 58	5	82	51 03	3	81	51 06	2	79	51 08	+1	77	51 09	-1
42	49 39	9	87	49 48	7	85	49 55	6	84	50 01	5	82	50 06	4	81	50 10	2	79	50 12	1	77	50 13	0
43	48 42	9	86	48 51	7	85	48 58	6	83	49 04	5	82	49 09	4	80	49 13	3	79	49 16	1	77	49 17	+1
44	47 45	8	86	47 53	8	85	48 01	6	83	48 07	6	82	48 13	4	80	48 17	3	79	48 20	2	77	48 22	0
45	46 47	+9	86	46 56	+8	85	47 04	+7	83	47 11	+5	82	47 16	+5	80	47 21	+3	79	47 24	+2	77	47 26	+1
46	45 50	9	86	45 59	8	84	46 07	7	83	46 14	6	82	46 20	4	80	46 24	4	79	46 28	2	77	46 30	1
47	44 53	9	86	45 02	8	84	45 10	7	83	45 17	6	81	45 23	5	80	45 28	4	79	45 32	3	77	45 35	2
48	43 56	9	85	44 05	8	84	44 13	7	83	44 20	7	81	44 27	5	80	44 32	4	78	44 36	3	77	44 39	2
49	42 59	9	85	43 08	8	84	43 16	8	82	43 24	6	81	43 30	6	80	43 36	4	78	43 40	4	77	43 44	2
50	42 01	+10	85	42 11	+8	84	42 19	+8	82	42 27	+7	81	42 34	+6	80	42 40	+4	78	42 44	+4	77	42 48	+3
51	41 04	10	85	41 14	9	83	41 23	7	82	41 30	7	81	41 37	6	79	41 43	5	78	41 48	5	77	41 53	3
52	40 07	10	85	40 17	9	83	40 26	8	82	40 34	7	81	40 41	6	79	40 47	6	78	40 53	4	77	40 57	4
53	39 10	10	84	39 20	9	83	39 29	8	82	39 37	8	80	39 45	6	79	39 51	6	78	39 57	5	77	40 02	4
54	38 13	10	84	38 23	9	83	38 32	9	82	38 41	7	80	38 48	7	79	38 55	6	78	39 01	5	76	39 06	4
55	37 16	+10	84	37 26	+10	83	37 36	+8	81	37 44	+8	80	37 52	+7	79	37 59	+6	78	38 05	+6	76	38 11	+4
56	36 19	10	84	36 29	10	82	36 39	9	81	36 48	8	80	36 56	7	79	37 03	7	77	37 10	5	76	37 15	5
57	35 22	10	83	35 32	10	82	35 42	9	81	35 51	8	80	35 59	6	79	36 07	7	77	36 14	6	76	36 20	5
58	34 25	11	83	34 36	9	82	34 45	10	81	34 55	8	80	35 03	8	78	35 11	7	77	35 18	6	76	35 24	6
59	33 28	11	83	33 39	10	82	33 49	9	81	33 58	9	79	34 07	8	78	34 15	8	77	34 23	6	76	34 29	6
60	32 31	+11	83	32 42	+10	82	32 52	+10	80	33 02	+9	79	33 11	+8	78	33 19	+8	77	33 27	+7	76	33 34	+6
61	31 34	11	83	31 45	11	81	31 56	9	80	32 06	9	79	32 15	8	78	32 23	8	77	32 31	8	76	32 39	6
62	30 37	11	82	30 48	11	81	30 59	10	80	31 09	10	79	31 19	9	78	31 28	8	77	31 36	7	75	31 43	7
63	29 40	12	82	29 52	11	81	30 03	10	80	30 13	10	79	30 23	9	78	30 32	8	76	30 40	8	75	30 48	7
64	28 44	11	82	28 55	11	81	29 06	11	80	29 17	10	79	29 27	9	77	29 36	9	76	29 45	8	75	29 53	8
65	27 47	+12	82	27 59	+11	81	28 10	+11	79	28 21	+10	78	28 31	+9	77	28 40	+9	76	28 49	+9	75	28 58	+8
66	26 50	12	82	27 02	11	80	27 13	11	79	27 24	11	78	27 35	10	77	27 45	9	76	27 54	9	75	28 03	8
67	25 53	12	81	26 05	12	80	26 17	11	79	26 28	11	78	26 39	10	77	26 49	10	76	26 59	9	75	27 08	8
68	24 57	12	81	25 09	12	80	25 21	11	79	25 32	11	78	25 43	10	77	25 53	10	76	26 03	10	75	26 13	9
69	24 00	12	81	24 12	12	80	24 24	12	79	24 36	11	78	24 47	11	77	24 58	10	75	25 08	10	74	25 18	9

15°	16°	17°	18°	19°	20°	21°	22°

S. Lat. {LHA greater than 180°........Zn=180−Z / LHA less than 180°...........Zn=180+Z

DECLINATION (15°-29°) SA

23° d	Z	24° Hc	d	Z	25° Hc	d	Z	26° Hc	d	Z	27° Hc	d	Z	28° Hc	d	Z	29° Hc	d	Z	LHA
-60	0	83 00	-60	0	82 00	-60	0	81 00	-60	0	80 00	-60	0	79 00	-60	0	78 00	-60	0	360
60	9	82 56	59	8	81 57	60	7	80 57	60	6	79 57	59	5	78 58	60	5	77 58	60	4	359
58	17	82 45	58	15	81 47	58	13	80 49	59	11	79 50	59	10	78 51	59	9	77 52	60	8	358
54	25	82 28	56	21	81 32	57	19	80 35	58	15	79 37	57	15	78 40	59	14	77 41	58	12	357
51	31	82 04	54	28	81 10	54	24	80 16	56	22	79 20	54	20	78 24	57	18	77 27	57	16	356
-48	37	81 35	-51	33	80 44	-52	29	79 52	-53	27	78 59	-55	24	78 04	-55	22	77 09	-56	20	355
44	42	81 02	48	38	80 14	49	34	79 25	51	31	78 34	53	28	77 41	53	26	76 48	54	24	354
41	47	80 25	44	42	79 41	47	38	78 54	49	35	78 05	50	32	77 15	52	29	76 23	52	27	353
38	50	79 45	46	46	79 04	44	42	78 20	46	38	77 34	48	35	76 46	50	32	75 56	51	30	352
35	53	79 03	39	49	78 24	41	45	77 43	44	41	76 59	46	38	76 13	47	36	75 26	49	33	351
-32	56	78 19	-36	52	77 43	-39	48	77 04	-41	44	76 23	-44	41	75 39	-45	38	74 54	-47	36	350
30	58	77 33	33	54	77 00	37	50	76 23	39	47	75 44	41	44	75 03	44	41	74 19	45	38	349
28	60	76 46	31	56	76 15	34	52	75 41	37	49	75 04	40	46	74 24	41	43	73 43	43	40	348
26	62	75 58	29	58	75 29	32	54	74 57	35	51	74 22	38	48	73 44	39	45	73 05	41	43	347
24	63	75 09	27	60	74 42	31	56	74 11	33	53	73 38	35	50	73 03	38	47	72 25	39	45	346
-23	65	74 19	-25	61	73 54	-29	58	73 25	-31	55	72 54	-34	52	72 20	-36	49	71 44	-37	46	345
21	66	73 29	24	62	73 05	27	59	72 38	30	56	72 08	31	53	71 37	35	51	71 02	36	48	344
19	67	72 38	23	63	72 15	25	60	71 50	28	57	71 22	30	55	70 52	33	52	70 19	34	49	343
18	68	71 46	21	65	71 25	24	62	71 01	26	59	70 35	29	56	70 06	31	53	69 35	32	51	342
17	68	70 54	20	65	70 34	22	63	70 12	25	60	69 47	27	57	69 20	30	55	68 50	31	52	341
-16	69	70 02	-19	66	69 43	-21	63	69 22	-23	61	68 59	-26	58	68 33	-28	56	68 05	-30	53	340
15	70	69 09	17	67	68 52	20	64	68 32	22	62	68 10	25	59	67 45	26	57	67 19	29	54	339
14	70	68 16	16	68	68 00	19	65	67 41	21	63	67 20	23	60	66 57	25	58	66 32	28	55	338
13	71	67 23	15	68	67 08	18	66	66 50	20	63	66 30	22	61	66 08	24	59	65 44	26	56	337
12	71	66 30	15	69	66 15	16	66	65 59	19	64	65 40	21	62	65 19	23	59	64 56	24	57	336
-11	72	65 36	-13	69	65 23	-16	67	65 07	-18	65	64 49	-19	62	64 30	-22	60	64 08	-24	58	335
11	72	64 42	12	70	64 30	15	67	64 15	17	65	63 58	18	63	63 40	21	61	63 19	22	59	334
9	72	63 49	12	70	63 37	14	68	63 23	16	66	63 07	18	64	62 49	19	61	62 30	21	59	333
9	73	62 55	11	70	62 44	13	68	62 31	15	66	62 16	17	64	61 59	19	62	61 40	20	60	332
8	73	62 01	11	71	61 50	12	69	61 38	14	67	61 24	16	65	61 08	17	63	60 51	20	61	331
-8	73	61 06	-9	71	60 57	-12	69	60 45	-13	67	60 32	-15	65	60 17	-16	63	60 01	-19	61	330
7	73	60 12	9	71	60 03	11	69	59 52	14	65	59 40	14	65	59 26	16	63	59 10	17	62	329
6	73	59 18	9	71	59 09	10	70	58 59	11	68	58 48	14	66	58 35	15	64	58 20	17	62	328
6	74	58 23	7	72	58 16	10	70	58 06	11	68	57 55	12	66	57 43	14	64	57 29	16	62	327
5	74	57 29	7	72	57 22	9	70	57 13	10	68	57 03	12	66	56 51	13	65	56 38	15	63	326
-5	74	56 34	-6	72	56 28	-8	70	56 20	-10	68	56 10	-11	67	55 59	-12	65	55 47	-14	63	325
4	74	55 40	6	72	55 34	8	70	55 26	6	69	55 18	11	67	55 07	11	65	54 56	14	64	324
4	74	54 45	5	72	54 40	7	70	54 33	8	69	54 25	10	67	54 15	11	66	54 04	12	64	323
4	74	53 50	4	72	53 46	7	71	53 39	7	69	53 32	9	67	53 23	10	66	53 13	12	64	322
2	74	52 56	5	73	52 51	5	71	52 46	7	69	52 39	8	68	52 31	10	66	52 21	11	64	321
-2	74	52 01	-4	73	51 57	-5	71	51 52	-6	69	51 46	-8	68	51 38	-9	66	51 29	-10	65	320
2	74	51 06	3	73	51 03	5	71	50 58	5	70	50 53	7	68	50 46	8	66	50 37	9	65	319
2	74	50 11	3	73	50 09	4	71	50 05	5	70	50 00	7	68	49 53	8	67	49 45	8	65	318
-1	74	49 17	3	73	49 14	3	71	49 11	5	70	49 06	6	68	49 00	7	67	48 53	8	65	317
0	74	48 22	2	73	48 20	3	71	48 17	4	70	48 13	5	68	48 08	7	67	48 01	7	65	316
0	74	47 27	-1	73	47 26	-3	71	47 23	-3	70	47 20	-5	68	47 15	-6	67	47 09	-7	65	315
0	74	46 32	1	73	46 31	1	71	46 29	3	70	46 26	4	68	46 22	5	67	46 17	6	66	314
0	74	45 37	0	73	45 37	2	71	45 35	2	70	45 33	4	69	45 29	4	67	45 25	6	66	313
+2	74	44 43	0	73	44 43	-1	71	44 42	1	70	44 40	3	69	44 37	5	67	44 32	5	66	312
2	74	43 48	0	73	43 48	0	71	43 48	1	71	43 40	2	69	43 44	4	67	43 40	4	66	311
+2	74	42 53	+1	73	42 54	0	71	42 54	-1	70	42 53	-2	69	42 51	-3	67	42 48	-4	66	310
1	74	41 58	1	73	41 59	+1	71	42 00	-1	70	41 59	1	69	41 58	1	67	41 55	3	66	309
2	74	41 03	2	73	41 05	1	71	41 06	0	70	41 06	-1	69	41 05	2	67	41 03	3	66	308
3	74	40 09	2	73	40 11	1	71	40 12	0	70	40 12	0	69	40 12	2	67	40 10	3	66	307
3	74	39 14	2	73	39 16	2	71	39 18	+1	70	39 19	0	69	39 19	-1	67	39 18	2	66	306
+4	74	38 19	+3	73	38 22	+2	71	38 24	+1	70	38 25	+1	69	38 26	0	67	38 26	-2	66	305
4	74	37 24	4	72	37 28	2	71	37 30	2	70	37 32	1	69	37 33	0	67	37 33	1	66	304
5	74	36 30	3	72	36 33	3	71	36 36	3	70	36 39	1	69	36 40	+1	67	36 41	-1	66	303
5	73	35 35	4	72	35 39	3	71	35 42	3	70	35 45	2	69	35 47	1	67	35 48	0	66	302
5	73	34 40	5	72	34 45	4	71	34 49	3	70	34 52	2	69	34 54	2	67	34 56	+1	66	301
+6	73	33 46	+5	72	33 51	+4	71	33 55	+3	70	33 58	+3	69	34 01	+2	67	34 03	+2	66	300
6	73	32 51	5	72	32 56	5	71	33 01	4	70	33 05	3	68	33 08	3	67	33 11	2	66	299
7	73	31 57	5	72	32 02	5	71	32 07	5	70	32 12	3	68	32 15	3	67	32 18	2	66	298
7	73	31 02	6	72	31 08	5	71	31 13	5	70	31 18	4	68	31 22	4	67	31 26	3	66	297
7	73	30 08	6	72	30 14	6	71	30 20	5	69	30 25	4	68	30 30	3	67	30 33	4	66	296
+7	73	29 13	+7	72	29 20	+6	70	29 26	+6	69	29 32	+5	68	29 37	+4	67	29 41	+4	66	295
8	73	28 19	7	71	28 26	6	70	28 32	6	69	28 38	6	68	28 44	5	67	28 49	4	66	294
8	72	27 24	8	71	27 32	7	70	27 39	6	69	27 45	6	68	27 51	5	67	27 56	5	66	293
8	72	26 30	8	71	26 38	7	70	26 45	7	69	26 52	6	68	26 58	6	67	27 04	5	66	292
9	72	25 36	8	71	25 44	8	70	25 52	7	69	25 59	7	68	26 06	6	67	26 12	6	66	291

23°	24°	25°	26°	27°	28°	29°	

LAT 17°

APPENDIX 5 (Part two   Problem three)

N. Lat. { LHA greater than 180°....... Zn=Z
{ LHA less than 180°.........Zn=360−Z

DECLINATION (15°-29) S.

LHA	15° Hc	d	Z	16° Hc	d	Z	17° Hc	d	Z	18° Hc	d	Z	19° Hc	d	Z	20° Hc	d	Z	21° Hc	d	Z	22° Hc
0	72 00	+60	180	73 00	+60	180	74 00	+60	180	75 00	+60	180	76 00	+60	180	77 00	+60	180	78 00	+60	180	79 00 +6
1	71 59	60	177	72 59	60	177	73 59	59	177	74 58	60	176	75 58	60	176	76 58	60	176	77 58	60	176	78 58 6
2	71 55	59	174	72 54	60	173	73 54	60	173	74 54	59	173	75 53	60	172	76 53	59	172	77 52	60	171	78 52 5
3	71 48	59	171	72 47	59	170	73 46	60	170	74 46	59	169	75 45	59	168	76 44	59	168	77 43	58	167	78 41 5
4	71 38	58	168	72 37	58	167	73 36	59	166	74 35	58	166	75 33	58	165	76 31	58	164	77 29	58	163	78 27 5
5	71 26	+59	165	72 25	+58	164	73 23	+58	163	74 21	+57	162	75 18	+57	161	76 15	+57	160	77 12	+57	158	78 09 +5
6	71 12	57	162	72 09	58	161	73 07	57	160	74 04	56	159	75 00	57	158	75 57	55	156	76 52	55	155	77 47 5
7	70 55	57	159	71 52	56	158	72 48	56	157	73 44	56	156	74 40	55	154	75 35	54	153	76 29	55	151	77 23 5
8	70 36	56	156	71 32	55	155	72 27	55	153	73 22	55	153	74 17	53	151	75 10	53	149	76 03	52	147	76 55 5
9	70 14	55	154	71 09	55	153	72 04	54	151	72 58	53	150	73 51	52	148	74 43	52	146	75 35	50	144	76 25 4
10	69 51	+54	151	70 45	+53	150	71 38	+53	148	72 31	+52	147	73 23	+51	145	74 14	+50	143	75 04	+49	141	75 53 +4
11	69 25	53	148	70 18	53	147	71 11	51	146	72 02	51	144	72 53	50	142	73 43	49	140	74 32	47	138	75 19 4
12	68 58	52	146	69 50	51	145	70 41	51	143	71 32	50	141	72 22	48	140	73 10	47	138	73 57	46	135	74 43 4
13	68 29	51	144	69 20	50	143	70 10	50	141	71 00	49	140	71 48	47	137	72 35	46	135	73 21	44	133	74 05 4
14	67 58	50	142	68 48	50	140	69 38	48	138	70 26	47	137	71 13	46	135	71 59	45	133	72 44	43	131	73 27 4
15	67 26	+49	139	68 15	+48	138	69 03	+48	136	69 51	+46	134	70 37	+44	133	71 21	+44	131	72 05	+41	128	72 46 +4
16	66 53	48	137	67 41	47	136	68 28	46	134	69 14	45	132	69 59	43	130	70 42	43	128	71 25	40	126	72 05 3
17	66 18	47	135	67 05	46	134	67 51	45	132	68 36	44	130	69 20	42	128	70 02	41	125	70 43	40	124	71 23 3
18	65 42	46	134	66 28	45	132	67 13	44	130	67 57	43	129	68 40	41	127	69 21	40	125	70 01	39	122	70 40 3
19	65 05	45	132	65 50	44	130	66 34	43	129	67 17	42	127	67 59	41	125	68 40	38	123	69 18	38	121	69 56 3
20	64 27	+44	130	65 11	+44	128	65 55	+42	127	66 37	+40	125	67 17	+40	123	67 57	+38	121	68 35	+36	119	69 11 +3
21	63 48	43	128	64 31	43	127	65 14	41	125	65 55	40	123	66 35	38	122	67 13	38	120	67 51	35	118	68 26 3
22	63 08	43	127	63 51	41	125	64 32	41	124	65 13	39	122	65 52	37	120	66 29	37	118	67 06	34	116	67 40 3
23	62 27	42	125	63 09	41	124	63 50	40	122	64 30	38	120	65 08	37	119	65 45	35	117	66 20	34	115	66 54 3
24	61 46	41	124	62 27	40	122	63 07	39	121	63 46	37	119	64 23	36	117	64 59	35	115	65 34	33	113	66 07 3
25	61 04	+40	123	61 44	+39	121	62 23	+38	119	63 01	+37	118	63 38	+36	116	64 14	+34	114	64 48	+32	112	65 20 +3
26	60 21	40	121	61 01	38	120	61 39	38	118	62 17	36	116	62 53	36	113	63 28	33	113	64 01	32	111	64 33 3
27	59 38	39	120	60 17	38	118	60 55	36	117	61 31	36	115	62 07	34	113	62 41	33	112	63 14	31	110	63 45 3
28	58 54	38	119	59 32	37	117	60 09	37	116	60 46	34	114	61 20	34	112	61 54	32	111	62 26	31	109	62 57 2
29	58 09	38	117	58 47	37	116	59 24	35	114	59 59	35	113	60 34	33	111	61 07	31	109	61 38	31	108	62 09 2
30	57 25	+37	116	58 02	+36	115	58 38	+35	113	59 13	+34	112	59 47	+32	110	60 19	+31	108	60 50	+30	107	61 20 +2
31	56 39	37	115	57 16	35	114	57 51	35	112	58 26	33	111	58 59	32	109	59 31	31	107	60 02	29	106	60 31 2
32	55 53	37	114	56 30	35	113	57 05	34	111	57 39	34	110	58 11	32	108	58 43	30	107	59 13	30	105	59 43 2
33	55 07	36	113	55 43	35	112	56 18	33	110	56 51	32	109	57 23	32	107	57 55	30	106	58 25	28	104	58 53 2
34	54 21	35	112	54 56	34	111	55 30	33	109	56 03	32	108	56 35	31	106	57 06	30	105	57 36	28	103	58 04 2
35	53 34	+35	111	54 09	+34	110	54 43	+32	108	55 15	+32	107	55 47	+30	105	56 17	+30	104	56 47	+28	102	57 15 +2
36	52 47	34	110	53 21	34	109	53 55	32	107	54 27	31	106	54 58	30	105	55 28	29	103	55 57	28	101	56 25 2
37	52 00	34	109	52 34	32	108	53 06	32	107	53 38	31	105	54 09	30	104	54 39	29	102	55 08	28	101	55 36 2
38	51 12	34	108	51 46	32	107	52 18	32	106	52 50	30	104	53 20	30	103	53 50	29	101	54 19	27	100	54 46 2
39	50 24	33	108	50 57	33	106	51 30	31	105	52 01	30	104	52 31	30	102	53 01	28	101	53 29	27	99	53 56 2
40	49 36	+33	107	50 09	+32	105	50 41	+31	104	51 12	+30	103	51 42	+29	101	52 11	+28	100	52 39	+27	98	53 06 +2
41	48 48	32	106	49 20	32	105	49 52	31	103	50 23	30	102	50 53	29	101	51 22	27	99	51 49	27	98	52 16 2
42	47 59	32	105	48 31	32	104	49 03	30	103	49 33	30	101	50 03	29	100	50 32	27	99	50 59	27	97	51 26 2
43	47 11	32	104	47 43	31	103	48 14	30	102	48 44	29	101	49 13	29	99	49 42	27	98	50 09	27	96	50 36 2
44	46 22	31	104	46 53	31	102	47 24	31	101	47 55	29	100	48 24	28	98	48 52	27	97	49 19	27	96	49 46 2
45	45 33	+31	103	46 04	+31	102	46 35	+30	100	47 05	+29	99	47 34	+28	98	48 02	+27	96	48 29	+27	95	48 56 +2
46	44 44	31	102	45 15	30	101	45 45	30	100	46 15	29	98	46 44	28	97	47 12	27	96	47 39	26	95	48 05 2
47	43 54	31	101	44 25	31	100	44 56	29	99	45 25	29	98	45 54	28	96	46 22	27	96	46 49	26	94	47 15 2
48	43 05	31	101	43 36	30	100	44 06	29	98	44 35	29	97	45 04	28	96	45 32	27	95	45 59	26	93	46 25 2
49	42 15	31	100	42 46	30	99	43 16	29	98	43 45	29	96	44 14	28	95	44 42	27	94	45 09	26	93	45 35 2
50	41 26	+30	99	41 56	+30	98	42 26	+29	97	42 55	+29	96	43 24	+27	95	43 51	+27	93	44 18	+26	92	44 44 +2
51	40 36	30	99	41 06	30	98	41 36	29	96	42 05	29	95	42 34	27	94	43 01	27	93	43 28	26	92	43 54 2
52	39 46	31	98	40 17	29	97	40 46	29	96	41 15	28	95	41 43	28	93	42 11	27	92	42 38	26	91	43 04 2
53	38 56	31	97	39 27	29	96	39 56	29	95	40 25	28	94	40 53	28	93	41 21	26	92	41 47	26	90	42 13 2
54	38 06	31	97	38 37	29	96	39 06	29	95	39 35	28	93	40 03	27	92	40 30	27	91	40 57	26	90	41 23 2
55	37 16	+30	96	37 46	+30	95	38 16	+29	94	38 45	+28	93	39 13	+27	92	39 40	+27	91	40 07	+26	89	40 33 +2
56	36 26	30	96	36 56	30	94	37 26	28	93	37 54	28	92	38 22	27	91	38 50	26	90	39 16	26	89	39 42 2
57	35 36	30	95	36 06	29	94	36 35	28	93	37 04	28	92	37 32	27	91	37 59	27	89	38 26	26	88	38 52 5
58	34 46	30	94	35 16	29	93	35 45	28	92	36 14	28	91	36 42	27	90	37 09	27	89	37 36	26	88	38 02 2
59	33 56	30	94	34 26	29	93	34 55	28	92	35 23	28	91	35 51	27	89	36 19	27	88	36 46	26	87	37 12 2
60	33 06	+29	93	33 35	+29	92	34 04	+29	91	34 33	+28	90	35 01	+27	89	35 28	+27	88	35 55	+27	87	36 22 +2
61	32 15	30	93	32 45	29	91	33 14	29	91	33 43	28	89	34 11	27	88	34 38	27	87	35 05	26	86	35 31 2
62	31 25	30	92	31 55	29	91	32 24	28	90	32 52	28	89	33 20	28	88	33 48	27	87	34 15	26	85	34 41 2
63	30 35	29	91	31 04	29	90	31 33	28	89	32 02	28	88	32 30	28	87	32 58	27	86	33 25	26	85	33 51 2
64	29 45	29	91	30 14	29	90	30 43	29	89	31 12	28	88	31 40	27	87	32 07	26	86	32 35	26	85	33 01 2
65	28 54	+30	90	29 24	+29	89	29 53	+28	88	30 21	+29	87	30 50	+27	86	31 17	+27	85	31 44	+27	84	32 11 +2
66	28 04	29	90	28 33	30	89	29 03	28	88	29 31	28	87	29 59	28	86	30 27	27	85	30 54	27	84	31 21 2
67	27 14	29	89	27 43	29	88	28 12	29	87	28 41	28	86	29 09	28	85	29 37	27	84	30 04	27	83	30 31 2
68	26 23	30	89	26 53	29	88	27 22	29	87	27 51	28	86	28 19	28	85	28 47	28	84	29 15	27	83	29 42 2
69	25 33	30	88	26 03	29	87	26 32	29	86	27 01	28	85	27 29	28	84	27 57	28	83	28 25	27	82	28 52 2

| | 15° | | | 16° | | | 17° | | | 18° | | | 19° | | | 20° | | | 21° | | | 22° |

S. Lat. { LHA greater than 180°.........Zn=180−Z
{ LHA less than 180°............Zn=180+Z

DECLINATION (15°-29°) SA

23°			24°			25°			26°			27°			28°			29°			LHA
Hc	d	Z	Hc	d	Z	Hc	d	Z	Hc	d	Z	Hc	d	Z	Hc	d	Z	Hc	d	Z	
	+60	180	81 00	+60	180	82 00	+60	180	83 00	+60	180	84 00	+60	180	85 00	+60	180	86 00	+60	180	360
	59	175	80 57	60	174	81 57	60	174	82 57	59	173	83 56	60	172	84 56	59	170	85 55	58	168	359
	59	170	80 50	59	169	81 49	58	167	82 47	58	166	83 45	58	163	84 43	56	161	85 39	54	156	358
	58	164	80 37	58	163	81 35	57	161	82 32	56	159	83 28	54	156	84 22	53	152	85 15	49	147	357
	56	160	80 20	56	158	81 16	55	155	82 11	53	153	83 04	52	149	83 56	48	144	84 44	44	138	356
	+55	155	79 59	+54	153	80 53	+53	150	81 46	+50	147	82 36	+48	143	83 24	+44	138	84 08	+39	132	355
	54	151	79 35	51	148	80 26	50	145	81 16	48	142	82 04	45	138	82 49	40	133	83 29	35	126	354
	51	147	79 06	50	144	79 56	48	141	80 44	45	137	81 29	41	133	82 10	38	128	82 48	31	122	353
	49	143	78 35	44	140	79 23	45	137	80 08	42	133	80 50	39	129	81 29	35	124	82 04	29	118	352
	48	139	78 02	45	136	78 47	43	133	79 30	40	130	80 10	37	125	80 47	32	121	81 19	27	115	351
	+45	136	77 26	+44	133	78 10	+41	130	78 51	+37	126	79 28	+35	122	80 03	+30	118	80 33	+25	112	350
	44	133	76 49	41	130	77 30	39	127	78 09	36	123	78 45	33	119	79 18	28	115	79 46	24	110	349
	42	130	76 09	40	128	76 49	38	124	77 27	36	122	78 01	31	117	78 32	26	113	78 58	23	108	348
	41	128	75 29	38	125	76 07	36	122	76 43	32	118	77 15	30	115	77 45	25	111	78 10	22	106	347
	39	126	74 47	37	123	75 24	34	120	75 58	31	116	76 29	28	113	76 57	25	109	77 22	20	105	346
	+38	123	74 04	+36	121	74 40	+33	118	75 13	+30	114	75 43	+27	111	76 10	+23	107	76 33	+20	103	345
	36	121	73 20	35	119	73 55	31	116	74 26	29	113	74 55	26	109	75 21	23	106	75 44	19	102	344
	36	119	72 36	33	117	73 09	31	114	73 40	29	110	74 08	25	108	74 33	22	104	74 55	18	101	343
	34	118	71 50	38	115	72 23	29	112	72 52	28	109	73 20	24	106	73 44	21	103	74 05	18	100	342
	34	116	71 05	31	114	71 36	29	111	72 05	26	108	72 31	24	105	72 55	20	102	73 15	18	99	341
	+32	115	70 18	+31	112	70 49	+28	109	71 17	+25	107	71 42	+23	104	72 05	+21	101	72 26	+17	98	340
	31	113	69 31	30	111	70 01	27	108	70 28	25	106	70 53	23	103	71 16	20	100	71 36	17	97	339
	31	112	68 44	29	109	69 13	27	107	69 40	24	104	70 04	22	102	70 26	20	99	70 46	16	96	338
	30	110	67 56	29	108	68 25	26	106	68 51	24	103	69 15	22	101	69 37	19	98	69 56	16	95	337
	29	109	67 08	28	107	67 36	26	105	68 02	23	102	68 25	22	100	68 47	19	97	69 06	16	95	336
	+29	108	66 20	+27	106	66 47	+26	104	67 13	+23	101	67 36	+21	99	67 57	+18	96	68 15	+17	94	335
	29	107	65 32	26	105	65 58	25	103	66 23	23	100	66 46	21	98	67 07	18	96	67 25	16	93	334
	28	106	64 43	26	104	65 09	25	102	65 34	22	100	65 56	21	97	66 17	18	95	66 35	16	93	333
	28	105	63 54	25	103	64 20	24	101	64 44	22	99	65 06	20	97	65 26	19	94	65 45	16	92	332
	27	104	63 05	25	102	63 30	24	100	63 54	22	98	64 16	20	96	64 36	18	94	64 54	16	91	331
	+26	103	62 15	+26	101	62 41	+23	99	63 04	+22	97	63 26	+20	95	63 46	+18	93	64 04	+16	91	330
	27	102	61 26	25	100	61 51	23	98	62 14	22	96	62 36	20	94	62 56	18	92	63 14	16	90	329
	26	101	60 36	25	100	61 01	23	98	61 24	22	96	61 46	19	94	62 05	18	92	62 23	16	90	328
	26	101	59 47	24	99	60 11	23	97	60 34	21	95	60 55	20	93	61 15	18	91	61 33	16	89	327
	26	100	58 57	24	98	59 21	23	96	59 44	21	94	60 05	20	93	60 25	18	91	60 43	16	89	326
	+25	99	58 07	+24	97	58 31	+23	96	58 54	+21	94	59 15	+19	92	59 34	+18	90	59 52	+17	88	325
	25	98	57 17	24	97	57 41	23	95	58 04	21	93	58 25	19	91	58 44	18	90	59 02	16	88	324
	25	98	56 27	24	96	56 51	22	94	57 13	21	93	57 34	20	91	57 54	18	89	58 12	16	87	323
	25	97	55 37	24	95	56 01	22	94	56 23	21	92	56 44	19	90	57 03	19	88	57 22	16	87	322
	25	96	54 47	23	95	55 10	23	93	55 33	21	91	55 54	19	90	56 13	18	88	56 31	17	86	321
	+25	95	53 57	+23	94	54 20	+22	92	54 42	+21	91	55 03	+20	89	55 23	+18	87	55 41	+17	86	320
	24	95	53 06	24	93	53 30	22	92	53 52	21	90	54 13	20	89	54 33	18	87	54 51	17	85	319
	24	94	52 16	24	93	52 40	22	91	53 02	21	90	53 23	20	88	53 42	19	87	54 01	17	85	318
	24	94	51 26	23	92	51 49	22	91	52 11	21	89	52 32	20	88	52 52	19	86	53 11	17	84	317
	25	93	50 36	23	92	50 59	22	90	51 21	21	89	51 42	20	87	52 02	19	86	52 21	17	84	316
	+24	92	49 45	+24	91	50 09	+22	90	50 31	+21	88	50 52	+20	87	51 12	+19	85	51 31	+17	84	315
	24	92	48 55	23	90	49 18	23	89	49 41	21	88	50 02	20	86	50 22	19	85	50 41	17	83	314
	25	91	48 05	23	90	48 28	22	89	48 50	22	87	49 12	20	86	49 32	19	84	49 51	18	83	313
	24	91	47 14	24	89	47 38	22	88	48 00	21	88	48 21	21	85	48 42	19	84	49 01	18	82	312
	24	90	46 24	23	89	46 47	23	87	47 10	21	86	47 31	21	85	47 52	19	83	48 11	18	82	311
	+25	90	45 34	+23	88	45 57	+23	87	46 20	+21	86	46 41	+21	84	47 02	+19	83	47 21	+19	82	310
	24	89	44 43	24	88	45 07	22	86	45 29	22	85	45 51	21	84	46 12	19	82	46 31	19	81	309
	24	89	43 53	24	87	44 17	22	86	44 39	22	85	45 01	21	83	45 22	20	82	45 42	19	81	308
	24	88	43 03	24	87	43 27	22	86	43 49	22	84	44 11	21	83	44 32	20	82	44 52	19	80	307
	25	88	42 13	23	86	42 36	23	85	42 59	22	84	43 21	21	83	43 42	21	81	44 03	19	80	306
	+25	87	41 23	+23	86	41 46	+23	85	42 09	+22	83	42 31	+22	82	42 53	+20	81	43 13	+20	80	305
	24	87	40 32	24	85	40 56	23	84	41 19	23	83	41 42	21	82	42 03	21	80	42 24	19	79	304
	24	86	39 42	24	85	40 06	23	84	40 29	23	83	40 52	21	81	41 13	21	80	41 34	20	79	303
	25	86	38 52	24	84	39 16	24	83	39 40	22	82	40 02	22	81	40 24	21	80	40 45	20	78	302
	25	85	38 02	24	84	38 26	24	83	38 50	22	82	39 12	23	80	39 35	21	79	39 56	20	78	301
	+25	85	37 12	+24	83	37 36	+24	82	38 00	+23	81	38 23	+22	80	38 45	+22	79	39 07	+20	78	300
	25	84	36 22	25	83	36 47	23	82	37 10	23	81	37 33	23	79	37 56	22	78	38 18	21	77	299
	25	84	35 32	25	82	35 57	23	81	36 21	23	80	36 44	23	79	37 07	22	78	37 29	21	77	298
	25	83	34 42	25	82	35 07	24	81	35 31	24	80	35 55	22	79	36 17	23	77	36 40	21	76	297
	26	83	33 53	25	82	34 17	25	80	34 42	23	79	35 05	23	78	35 28	23	77	35 51	22	76	296
	+26	82	33 03	+25	81	33 28	+24	80	33 52	+24	79	34 16	+23	79	34 39	+23	77	35 02	+22	76	295
	26	82	32 13	25	81	32 38	25	80	33 03	24	78	33 27	24	77	33 50	23	76	34 13	23	75	294
	26	81	31 24	25	80	31 49	25	79	32 14	24	78	32 38	24	77	33 02	23	76	33 25	22	75	293
	26	81	30 34	26	80	31 00	25	79	31 25	24	78	31 49	24	77	32 13	23	75	32 36	23	74	292
	27	80	29 45	25	79	30 10	25	78	30 35	25	77	31 00	24	76	31 24	24	75	31 48	23	74	291

23°	24°	25°	26°	27°	28°	29°

NAME AS LATITUDE

LAT 33°

LAT 33°

TABLE 5.—Correction to Tabula...

d / r	1	2	3	4	5	6	7	8	9	10	11	12	13	14	15	16	17	18	19	20	21	22	23	24	25	26	27	28	29
0	0	0	0	0	0	0	0	0	0	0	0	0	0	0	0	0	0	0	0	0	0	0	0	0	0	0	0	0	
1	0	0	0	0	0	0	0	0	0	0	0	0	0	0	0	0	0	0	0	0	0	0	0	0	0	0	0	0	
2	0	0	0	0	0	0	0	0	0	0	0	0	0	0	0	1	1	1	1	1	1	1	1	1	1	1	1	1	
3	0	0	0	0	0	0	0	0	0	0	1	1	1	1	1	1	1	1	1	1	1	1	1	1	1	1	1	1	
4	0	0	0	0	0	0	0	1	1	1	1	1	1	1	1	1	1	1	1	1	1	1	2	2	2	2	2	2	
5	0	0	0	0	0	0	1	1	1	1	1	1	1	1	1	1	1	2	2	2	2	2	2	2	2	2	2	2	
6	0	0	0	0	0	1	1	1	1	1	1	1	1	1	2	2	2	2	2	2	2	2	2	2	2	3	3	3	
7	0	0	0	0	1	1	1	1	1	1	1	1	2	2	2	2	2	2	2	2	2	3	3	3	3	3	3	3	
8	0	0	0	1	1	1	1	1	1	1	1	2	2	2	2	2	2	2	3	3	3	3	3	3	3	3	4	4	
9	0	0	0	1	1	1	1	1	1	2	2	2	2	2	2	2	3	3	3	3	3	3	3	4	4	4	4	4	
10	0	0	0	1	1	1	1	1	2	2	2	2	2	2	2	3	3	3	3	3	4	4	4	4	4	4	4	5	
11	0	0	1	1	1	1	1	1	2	2	2	2	2	3	3	3	3	3	3	4	4	4	4	4	5	5	5	5	
12	0	0	1	1	1	1	1	2	2	2	2	2	3	3	3	3	3	4	4	4	4	4	5	5	5	5	5	6	
13	0	0	1	1	1	1	2	2	2	2	2	3	3	3	3	3	4	4	4	4	5	5	5	5	5	6	6	6	
14	0	0	1	1	1	1	2	2	2	2	3	3	3	3	4	4	4	4	4	5	5	5	5	6	6	6	6	7	
15	0	0	1	1	1	2	2	2	2	2	3	3	3	4	4	4	4	4	5	5	5	6	6	6	6	6	7	7	
16	0	1	1	1	1	2	2	2	2	3	3	3	3	4	4	4	5	5	5	5	6	6	6	6	7	7	7	7	
17	0	1	1	1	1	2	2	2	3	3	3	3	4	4	4	5	5	5	5	6	6	6	7	7	7	7	8	8	
18	0	1	1	1	2	2	2	2	3	3	3	4	4	4	4	5	5	5	6	6	6	7	7	7	8	8	8	8	
19	0	1	1	1	2	2	2	3	3	3	3	4	4	4	5	5	5	6	6	6	7	7	7	8	8	8	9	9	
20	0	1	1	1	2	2	2	3	3	3	4	4	4	5	5	5	6	6	6	7	7	7	8	8	8	9	9	9	1
21	0	1	1	1	2	2	2	3	3	4	4	4	5	5	5	6	6	6	7	7	7	8	8	8	9	9	9	10	1
22	0	1	1	1	2	2	3	3	3	4	4	4	5	5	6	6	6	7	7	7	8	8	8	9	9	10	10	10	1
23	0	1	1	2	2	2	3	3	3	4	4	5	5	5	6	6	7	7	7	8	8	8	9	9	10	10	10	11	1
24	0	1	1	2	2	2	3	3	4	4	4	5	5	6	6	6	7	7	8	8	8	9	9	10	10	10	11	11	1
25	0	1	1	2	2	2	3	3	4	4	5	5	5	6	6	7	7	8	8	8	9	9	10	10	11	11	11	12	1
26	0	1	1	2	2	3	3	3	4	4	5	5	6	6	6	7	7	8	8	9	9	10	10	11	11	11	12	12	1
27	0	1	1	2	2	3	3	4	4	4	5	5	6	6	7	7	8	8	9	9	9	10	10	11	11	12	12	13	1
28	0	1	1	2	2	3	3	4	4	5	5	6	6	7	7	7	8	8	9	9	10	10	11	11	12	12	13	13	1
29	0	1	1	2	2	3	3	4	4	5	5	6	6	7	7	8	8	9	9	10	10	11	11	12	12	13	13	14	1
30	0	1	2	2	2	3	4	4	4	5	6	6	6	7	8	8	8	9	10	10	10	11	12	12	12	13	14	14	1
31	1	1	2	2	3	3	4	4	5	5	6	6	7	7	8	8	9	9	10	10	11	11	12	12	13	13	14	14	1
32	1	1	2	2	3	3	4	4	5	5	6	6	7	7	8	9	9	10	10	11	11	12	12	13	13	14	14	15	1
33	1	1	2	2	3	3	4	4	5	6	6	7	7	8	8	9	9	10	10	11	12	12	13	13	14	14	15	15	1
34	1	1	2	2	3	3	4	5	5	6	6	7	7	8	8	9	10	10	11	11	12	12	13	14	14	15	15	16	1
35	1	1	2	2	3	4	4	5	5	6	6	7	8	8	9	9	10	10	11	12	12	13	13	14	15	15	16	16	1
36	1	1	2	2	3	4	4	5	5	6	7	7	8	8	9	10	10	11	11	12	13	13	14	14	15	16	16	17	1
37	1	1	2	2	3	4	4	5	6	6	7	7	8	9	9	10	10	11	12	12	13	14	14	15	15	16	17	17	1
38	1	1	2	3	3	4	4	5	6	6	7	8	8	9	10	10	11	11	12	13	13	14	15	15	16	16	17	18	1
39	1	1	2	3	3	4	5	5	6	6	7	8	8	9	10	10	11	12	12	13	14	14	15	16	16	17	18	18	1
40	1	1	2	3	3	4	5	5	6	7	7	8	9	9	10	11	11	12	13	13	14	15	15	16	17	17	18	19	1
41	1	1	2	3	3	4	5	5	6	7	8	8	9	10	10	11	12	12	13	14	14	15	16	16	17	18	18	19	2
42	1	1	2	3	4	4	5	6	6	7	8	8	9	10	10	11	12	13	13	14	15	15	16	17	18	18	19	20	2
43	1	1	2	3	4	4	5	6	6	7	8	9	9	10	11	11	12	13	14	14	15	16	16	17	18	19	19	20	2
44	1	1	2	3	4	4	5	6	7	7	8	9	10	10	11	12	12	13	14	15	15	16	17	18	18	19	20	21	2
45	1	2	2	3	4	4	5	6	7	8	8	9	10	10	11	12	13	14	14	15	16	16	17	18	19	20	20	21	2
46	1	2	2	3	4	5	5	6	7	8	8	9	10	11	12	12	13	14	15	15	16	17	18	18	19	20	21	21	2
47	1	2	2	3	4	5	5	6	7	8	9	9	10	11	12	13	13	14	15	16	16	17	18	19	20	20	21	22	2
48	1	2	2	3	4	5	6	6	7	8	9	10	10	11	12	13	14	14	15	16	17	18	18	19	20	21	22	22	2
49	1	2	2	3	4	5	6	7	7	8	9	10	11	11	12	13	14	15	16	16	17	18	19	20	20	21	22	23	2
50	1	2	2	3	4	5	6	7	8	8	9	10	11	12	12	13	14	15	16	17	18	18	19	20	21	22	22	23	2
51	1	2	3	3	4	5	6	7	8	8	9	10	11	12	13	14	14	15	16	17	18	19	20	20	21	22	23	24	2
52	1	2	3	3	4	5	6	7	8	9	10	10	11	12	13	14	15	16	16	17	18	19	20	21	22	23	23	24	2
53	1	2	3	4	4	5	6	7	8	9	10	11	11	12	13	14	15	16	17	18	19	20	21	22	22	23	24	25	2
54	1	2	3	4	4	5	6	7	8	9	10	11	12	13	14	14	15	16	17	18	19	20	21	22	22	23	24	25	2
55	1	2	3	4	5	6	6	7	8	9	10	11	12	13	14	15	16	16	17	18	19	20	21	22	23	24	25	26	2
56	1	2	3	4	5	6	7	7	8	9	10	11	12	13	14	15	16	17	18	19	20	21	21	22	23	24	25	26	2
57	1	2	3	4	5	6	7	8	9	10	10	11	12	13	14	15	16	17	18	19	20	21	22	23	24	25	26	27	28
58	1	2	3	4	5	6	7	8	9	10	11	12	13	14	14	15	16	17	18	19	20	21	22	23	24	25	26	27	28
59	1	2	3	4	5	6	7	8	9	10	11	12	13	14	15	16	17	18	19	20	21	22	23	24	25	26	27	28	29

32	33	34	35	36	37	38	39	40	41	42	43	44	45	46	47	48	49	50	51	52	53	54	55	56	57	58	59	60	d/'
0	0	0	0	0	0	0	0	0	0	0	0	0	0	0	0	0	0	0	0	0	0	0	0	0	0	0	0	0	0
1	1	1	1	1	1	1	1	1	1	1	1	1	1	1	1	1	1	1	1	1	1	1	1	1	1	1	1	1	1
1	1	1	1	1	1	1	1	1	1	1	1	1	2	2	2	2	2	2	2	2	2	2	2	2	2	2	2	2	2
2	2	2	2	2	2	2	2	2	2	2	2	2	2	2	2	2	2	2	3	3	3	3	3	3	3	3	3	3	3
2	2	2	2	2	2	3	3	3	3	3	3	3	3	3	3	3	3	3	3	3	4	4	4	4	4	4	4	4	4
3	3	3	3	3	3	3	3	3	3	4	4	4	4	4	4	4	4	4	4	4	4	4	5	5	5	5	5	5	5
3	3	3	4	4	4	4	4	4	4	4	4	4	4	5	5	5	5	5	5	5	5	5	6	6	6	6	6	6	6
4	4	4	4	4	4	4	5	5	5	5	5	5	5	5	5	6	6	6	6	6	6	6	6	7	7	7	7	7	7
4	4	5	5	5	5	5	5	5	5	6	6	6	6	6	6	6	7	7	7	7	7	7	7	7	8	8	8	8	8
5	5	5	5	5	6	6	6	6	6	6	6	7	7	7	7	7	7	8	8	8	8	8	8	8	9	9	9	9	9
5	6	6	6	6	6	6	6	7	7	7	7	7	8	8	8	8	8	8	8	9	9	9	9	9	10	10	10	10	10
6	6	6	6	7	7	7	7	7	8	8	8	8	8	8	9	9	9	9	9	10	10	10	10	10	10	11	11	11	11
6	7	7	7	7	7	8	8	8	8	8	9	9	9	9	9	10	10	10	10	10	11	11	11	11	11	12	12	12	12
7	7	7	8	8	8	8	8	9	9	9	9	10	10	10	10	10	11	11	11	11	11	12	12	12	12	13	13	13	13
7	8	8	8	8	9	9	9	9	10	10	10	10	10	11	11	11	11	12	12	12	12	13	13	13	13	14	14	14	14
8	8	8	9	9	9	10	10	10	10	10	11	11	11	12	12	12	12	12	13	13	13	14	14	14	14	14	15	15	15
9	9	9	9	10	10	10	10	11	11	11	11	12	12	12	13	13	13	13	14	14	14	14	15	15	15	15	16	16	16
9	9	10	10	10	10	11	11	11	12	12	12	12	13	13	13	14	14	14	14	15	15	15	16	16	16	16	17	17	17
10	10	10	10	11	11	11	12	12	12	13	13	13	14	14	14	14	15	15	15	16	16	16	16	17	17	17	18	18	18
10	10	11	11	11	12	12	12	13	13	13	14	14	14	15	15	15	16	16	16	16	17	17	17	18	18	18	19	19	19
11	11	11	12	12	12	13	13	13	14	14	14	15	15	15	16	16	16	17	17	17	18	18	18	19	19	19	20	20	20
11	12	12	12	13	13	13	14	14	14	15	15	15	16	16	16	17	17	18	18	18	19	19	19	20	20	20	21	21	21
12	12	12	13	13	14	14	14	15	15	15	16	16	16	17	17	18	18	18	19	19	19	20	20	21	21	21	22	22	22
12	13	13	13	14	14	15	15	15	16	16	16	17	17	18	18	18	19	19	20	20	20	21	21	21	22	22	23	23	23
13	13	14	14	14	15	15	16	16	16	17	17	18	18	18	19	19	20	20	20	21	21	22	22	22	23	23	24	24	24
13	14	14	15	15	15	16	16	17	17	18	18	18	19	19	20	20	20	21	21	22	22	22	23	23	24	24	25	25	25
14	14	15	15	16	16	16	17	17	18	18	19	19	20	20	20	21	21	22	22	23	23	23	24	24	25	25	26	26	26
14	15	15	16	16	17	17	18	18	18	19	19	20	20	21	21	22	22	22	23	23	24	24	25	25	26	26	27	27	27
15	15	16	16	17	17	18	18	19	19	20	20	21	21	21	22	22	23	23	24	24	25	25	26	26	27	27	28	28	28
15	16	16	17	17	18	18	19	19	20	20	21	21	22	22	23	23	24	24	25	25	26	26	27	27	28	28	29	29	29
16	16	17	18	18	18	19	20	20	20	21	22	22	22	23	24	24	24	25	26	26	26	27	28	28	28	29	30	30	30
17	17	18	18	19	19	20	20	21	21	22	22	23	23	24	24	25	25	26	26	27	27	28	28	29	29	30	30	31	31
17	18	18	19	19	20	20	21	21	22	22	23	23	24	25	25	26	26	27	27	28	28	29	29	30	30	31	31	32	32
18	18	19	19	20	20	21	21	22	23	23	24	24	25	25	26	26	27	28	28	29	29	30	30	31	31	32	32	33	33
18	19	19	20	20	21	22	22	23	23	24	24	25	26	26	27	27	28	28	29	29	30	31	31	32	32	33	33	34	34
19	19	20	20	21	22	22	23	23	24	24	25	26	26	27	27	28	29	29	30	30	31	32	32	33	33	34	34	35	35
19	20	20	21	22	22	23	23	24	25	25	26	26	27	28	28	29	29	30	31	31	32	32	33	34	34	35	35	36	36
20	20	21	22	22	23	23	24	25	25	26	27	27	28	28	29	30	30	31	31	32	33	33	34	35	35	36	36	37	37
20	21	22	22	23	23	24	25	25	26	27	27	28	28	29	30	30	31	32	32	33	34	34	35	35	36	37	37	38	38
21	21	22	23	23	24	25	25	26	27	27	28	29	29	30	31	31	32	32	33	34	34	35	36	36	37	38	38	39	39
21	22	23	23	24	25	25	26	27	27	28	29	29	30	31	31	32	33	33	34	35	35	36	37	37	38	39	39	40	40
22	23	23	24	25	25	26	27	27	28	29	29	30	31	31	32	33	33	34	35	36	36	37	38	38	39	40	40	41	41
22	23	24	24	25	26	27	27	28	29	29	30	31	32	32	33	34	34	35	36	36	37	38	38	39	40	41	41	42	42
23	24	24	25	26	27	27	28	29	29	30	31	32	32	33	34	34	35	36	37	37	38	39	39	40	41	42	42	43	43
23	24	25	26	26	27	28	29	29	30	31	32	32	33	34	34	35	36	37	37	38	39	40	40	41	42	43	43	44	44
24	25	26	26	27	28	28	29	30	31	32	32	33	34	34	35	36	37	38	38	39	40	40	41	42	43	44	44	45	45
25	25	26	27	28	28	29	30	31	31	32	33	34	34	35	36	37	38	38	39	40	41	41	42	43	44	44	45	46	46
25	26	27	27	28	29	30	31	31	32	33	34	34	35	36	37	38	38	39	40	41	42	42	43	44	45	45	46	47	47
26	26	27	28	29	30	30	31	32	33	34	34	35	36	37	38	38	39	40	41	42	42	43	44	45	46	46	47	48	48
26	27	28	29	29	30	31	32	33	33	34	35	36	37	38	38	39	40	41	42	42	43	44	45	46	47	47	48	49	49
27	28	28	29	30	31	32	32	33	34	35	36	37	38	38	39	40	41	42	42	43	44	45	46	47	48	48	49	50	50
27	28	29	30	31	31	32	33	34	35	36	37	37	38	39	40	41	42	42	43	44	45	46	47	48	48	49	50	51	51
28	29	29	30	31	32	33	34	35	36	36	37	38	39	40	41	42	42	43	44	45	46	47	48	49	49	50	51	52	52
28	29	30	31	32	33	34	34	35	36	37	38	39	40	41	42	42	43	44	45	46	47	48	49	49	50	51	52	53	53
29	30	31	32	32	33	34	35	36	37	38	39	40	40	41	42	43	44	45	46	47	48	49	50	50	51	52	53	54	54
29	30	31	32	33	34	35	36	37	38	38	39	40	41	42	43	44	45	46	47	48	49	50	50	51	52	53	54	55	55
30	31	32	33	34	35	35	36	37	38	39	40	41	42	43	44	45	46	47	48	49	49	50	51	52	53	54	55	56	56
30	31	32	33	34	35	36	37	38	39	40	41	42	43	44	45	46	47	48	48	49	50	51	52	53	54	55	56	57	57
31	32	33	34	35	36	37	38	39	40	41	42	43	44	44	45	46	47	48	49	50	51	52	53	54	55	56	57	58	58
31	32	33	34	35	36	37	38	39	40	41	42	43	44	45	46	47	48	49	50	51	52	53	54	55	56	57	58	59	59

97

Sun Sight
Work Form

Date: _____   DR. Lat: _____   DR. Long: _____

Watch Time: _____   Height of Eye _____   HS: _____   IE: _____

Watch Error _____

Step 1

HS _____

IE (+ or −) _____

DIP (−) _____

App. Alt. _____

Alt. Corr. = _____

HO =
(HO = App. Alt. ± Alt. Corr.)

Step 2

Watch Time = _____

Watch Error ± = _____

Zone Time = _____

Zone Description = _____

GMT = _____

*(For daylight-savings time subtract one hour from Z D)

Step 3 (From Nautical Almanac)

GHA—Hours _____

GHA—M&S _____

Total GHA _____

Asmd. Long. _____

LHA* _____

Asmd. Lat. _____

Decl. N or S _____

Step 4 (From HO 249)

Z _____

ZN _____

d(+ or −) _____

Tab HC _____

Corr. for d _____

HC _____

HO _____

Intercept, T or A _____

*Whole degree only
LHA = GHa + E long, − W long.

APPENDIX 8

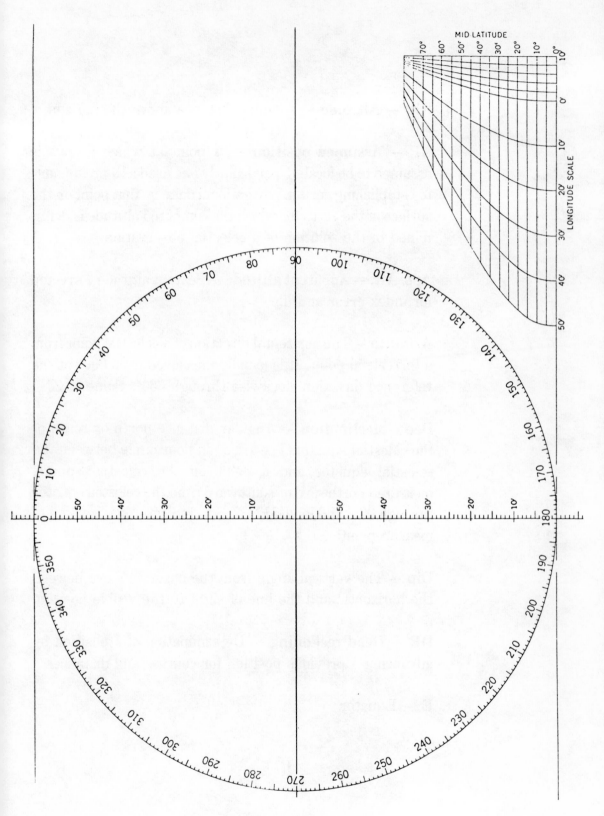

Universal Plotting Sheet

GLOSSARY

Alt. — Altitude — Angular distance above the horizon.

AP — Assumed position — A point at which a craft is assumed to be located, particularly one used as a preliminary to establishing certain navigational data, as that point on the surface of the earth for which the computed altitude is determined in the solution of a celestial observation.

App. Alt. — Apparent altitude — Sextant altitude corrected for index error and dip.

Azimuth — The horizontal direction of a celestial point from a terrestrial point. It is usually measured from 000° at the reference direction clockwise through 360°. Same as ZN.

Dec — declination — Angular distance north or south of the celestial equator. The arc of an hour circle between the celestial equator and a point on the celestial sphere, measured northward or southward from the celestial equator through 90°, and labeled N or S to indicate the direction of measurement.

Dip — The vertical angle from the observer's eye beween the horizontal and the line of sight to the visible horizon.

DR — Dead reckoning — Determination of a position by advancing a previous position for courses and distances.

E — Equator.

EP — Estimated position — The most probable position of a craft, determined from incomplete data or data of questionable accuracy.

Equation of time — The difference between the average time of meridian passage and the precise moment when it occurs on a particular day.

Fix — A relatively accurate position determined without reference to any former position.

GHA — Greenwich hour angle — Local hour angle at the Greenwich meridian.

GMT — Greenwich mean time — Local mean time at the Greenwich meridian.

GP — Geographical position — Point on earth's surface in line with center of earth and celestial body.

HC — Height computed — Altitude as determined using an assumed position and reduced according to the method described in Pub. No. 249.

HO — Height observed — The corrected sextant reading.

HS — Height shot — Uncorrected sextant reading.

IE — Instrument Error.

Intercept — Difference between HO and HC.

International date line — The boundary between the (–)12 and (+)12 time zones, corresponding approximately with the 180th meridian.

LAN — Local apparent noon.

Lat. — Latitude — Angular distance north or south of the equator; the arc of a meridian between the equator and a point on the surface of the earth, measured northward or southward from the equator through 90°, and labeled N or S to indicate the direction of measurement.

LHA — Local hour angle, angle formed by the chord from observer's location to the center of the earth and hence to a celestial object.

Long. — Longitude — Angular distance east or west of the prime, or Greenwich, meridian measured eastward or westward through 180° and labeled E or W to indicate the direction of measurement.

LOP — Line of position — A straight line somewhere on which the observer is located.

Meridian passage — Same as LAN; the moment of the sun's highest ascendency; when the sun is on the same meridian as the observer.

Refraction — The bending of a ray of light as it enters earth's atmosphere.

Running fix — A position determined by crossing lines of position with an appreciable time difference between them and advanced or retired to a common time.

ZN — Same as Azimuth, qv.

1. Take Dec. from daily pages of *Nautical Almanac* for whole hour nearest time of LAN.

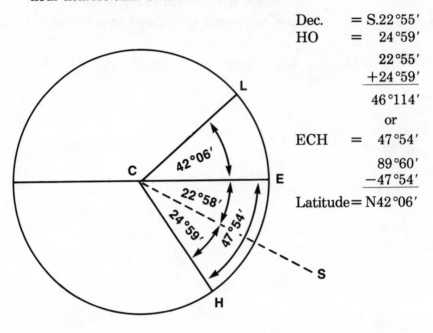

Dec.   = S.22°55'

HO   =   24°59'

          22°55'

        +24°59'

          46°114'

           or

ECH   =   47°54'

          89°60'

        −47°54'

Latitude= N42°06'

2.

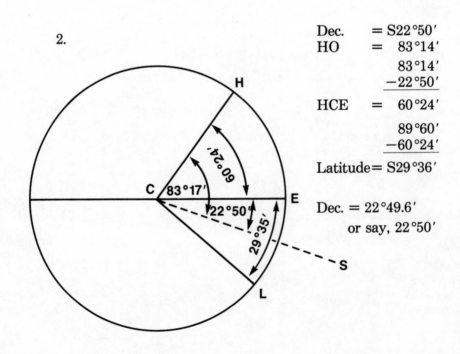

Dec.   = S22°50'

HO   =   83°14'

          83°14'

        −22°50'

HCE   =   60°24'

          89°60'

        −60°24'

Latitude= S29°36'

Dec. = 22°49.6'

   or say, 22°50'

1. Zone Time   12:10:05
    Zone Description +   8   (Add ZD for West
           GMT   20:10:05   Long., Subtract ZD
                                             for East Long.)

GHA of 20:00:00 hrs.= 118°57.6'   (whole hour)
                    +   2°31.3'   (increment for
                                10^m05^s from table)
              W 121°28.9'
          Say W 121°29'

2. ZT=   11:59:22
   ZD= −   4
GMT=   7:59:22

GHA of 7:00:00 hrs. = 284°08.4'   (whole hour)
                +   14°50.5'   (increment for
                               59^m22^s)
   TOTAL GHA = 298°58.9'   = 299°02.5'
                  359°60.0'
             − 298°58.9'
             E61°01.1'

To obtain East long. from GHA over 180°, subtract GHA
from 360°.

Problem #1 Solution

Sun Sight
Work Form

Date: July 4, 1982   DR. Lat: N 36° 51'   DR. Long: W 71° 20'

Watch Time: 07 32 17   Height of Eye 10'   HS: 21° 08'   IE: −03'

Watch Error   +12"

Step 1

HS	21° 08'
IE (+ or −)	+ 03'
DIP (−)	−03.1'
App. Alt.	21° 07.9'
Alt. Corr. =	+13.5'
	21° 21.4'

HO =
(HO = App. Alt. ± Alt. Corr.)

Step 2

Watch Time	=	07 32 17
Watch Error ±	=	− 12
Zone Time	=	07 32 05
Zone Description =		+ 4
GMT =		11 32 05

*(For daylight-savings time subtract
one hour from Z D)

Step 3 (From Nautical Almanac)

GHA—Hours	343° 55.6'
GHA—M&S	08° 01.3'
Total GHA (±360°)	351° 56.9'
Asmd. Long.	W 71° 56.9'
LHA*	280°
Asmd. Lat.	N 37°
Decl. N or S	N 22° 53.2'

Step 4 (From HO 249)

Z	78°
ZN	78°
d(+ or −)	+ 32'
Tab HC	20° 44'
Corr. for d	+28'
HC	21° 12'
HO	21° 24.4'
Intercept, T or A	T 12'.4

*Whole degree only
LHA = GHa + E long, − W long.

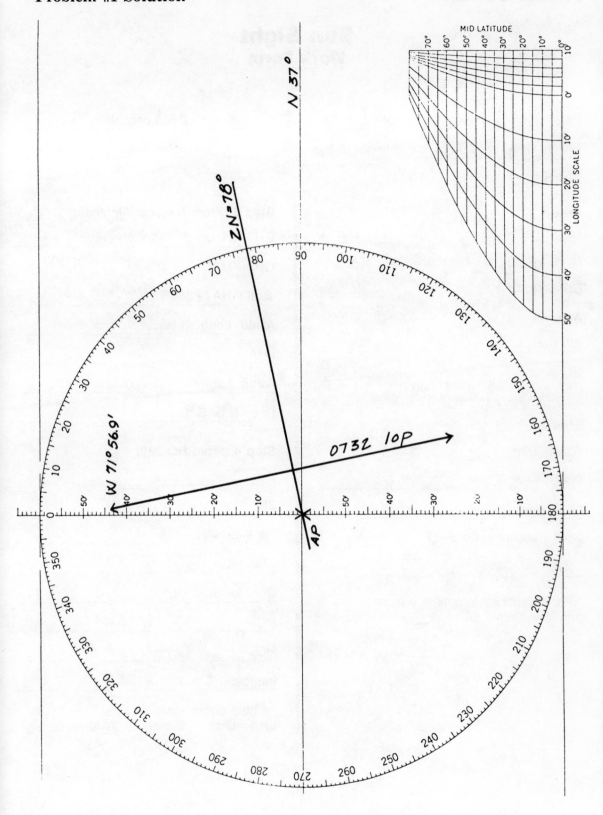

Problem #2 Solution

Sun Sight
Work Form

Date: Jan. 2, 1982   DR. Lat: S 17° 15'   DR. Long: W 148° 47'

Watch Time: 15 23 44   Height of Eye 10'   HS: 41° 27.2'   IE: +04'

Watch Error   -05

Step 1

HS	41° 27.2'
IE (+ or −)	−04'
DIP (−)	−03.1'
App. Alt.	41° 20.1'
Alt. Corr. =	+15.2'
	41° 35.3'

HO =
(HO = App. Alt. ± Alt. Corr.)

Step 2

Watch Time	=	15 23 44
Watch Error ±	=	+ 05
Zone Time	=	15 23 49
Zone Description*	=	+10
GMT = Jan. 3		25 23 49
		01 23 49

*(For daylight-savings time subtract one hour from Z D)

```
   75
+180
  255
```

Step 3 (From Nautical Almanac)

GHA—Hours	193° 56.1'
GHA—M&S	+ 5° 57.3'
Total GHA (±360°)	199° 53.4'
Asmd. Long.	(W 148° 53.4')
LHA*	51°
Asmd. Lat.	(S 17°)
Decl. N or S	S 22° 52.2'

Step 4 (From HO 249)

Z	75°
ZN	(255°)
d(+ or −)	+3'
Tab HC	41° 53'
Corr. for d	+3'
HC	41° 56'
HO	−41° 35.3'
Intercept, T or A	(A 20.7')

*Whole degree only
LHA = GHa + E long, − W long.

108

Problem #2 Solution

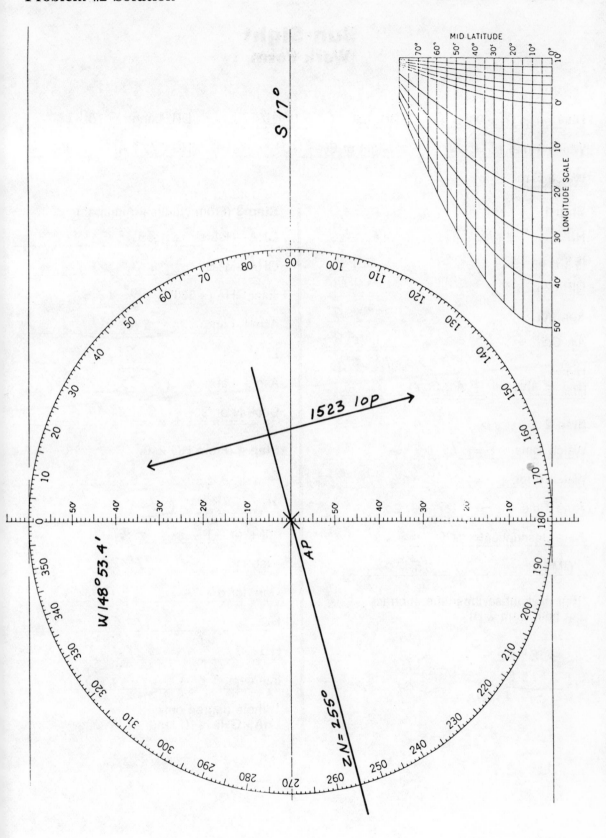

S 17°

MID LATITUDE

LONGITUDE SCALE

W 148° 53.4'

1523 10p

AP

Zn = 255°

109

Problem #3 Solution

Sun Sight
Work Form

Date: Jan. 3, 1982

Watch Time: 13 05 26

Watch Error +06

DR. Lat: S 32° 43'

Height of Eye 10'

DR. Long: E 16° 22'

HS: 72° 41'   IE: +06'

Step 1

HS	72° 41'
IE (+ or −)	−06'
DIP (−)	−03.1'
App. Alt.	72° 31.9'
Alt. Corr. =	+15.9'
HO =	72° 47.8'

(HO = App. Alt. ± Alt. Corr.)

Step 2

Watch Time	=	13 05 26
Watch Error ±	=	−06
Zone Time	=	13 05 20
Zone Description	=	−01
GMT =		12 05 20

*(For daylight-savings time subtract one hour from Z D)

GHA
358° 52.9'
+1° 20'
─────
359° 72.9'
OR
360° 12.9'
OR
00° 12.9'

180
+124
─────
304

Step 3 (From Nautical Almanac)

GHA—Hours	358° 52.9'
GHA—M&S	1° 20'
Total GHA (± 360°)	00° 12.9'
Asmd. Long.	(+ E 16° 47.1')
LHA*	17°
Asmd. Lat.	(S 33°)
Decl. N or S	S 22° 49.6'

Step 4 (From HO 249)

Z	122°
ZN	(302°)
d(+ or −)	+ 37'
Tab HC	71° 23'
Corr. for d	+ 30
HC	71° 53'
HO	72° 47.8'
Intercept, T or A	(T 54.8)

*Whole degree only
LHA = GHa + E long, − W long.

Problem #3 Solution

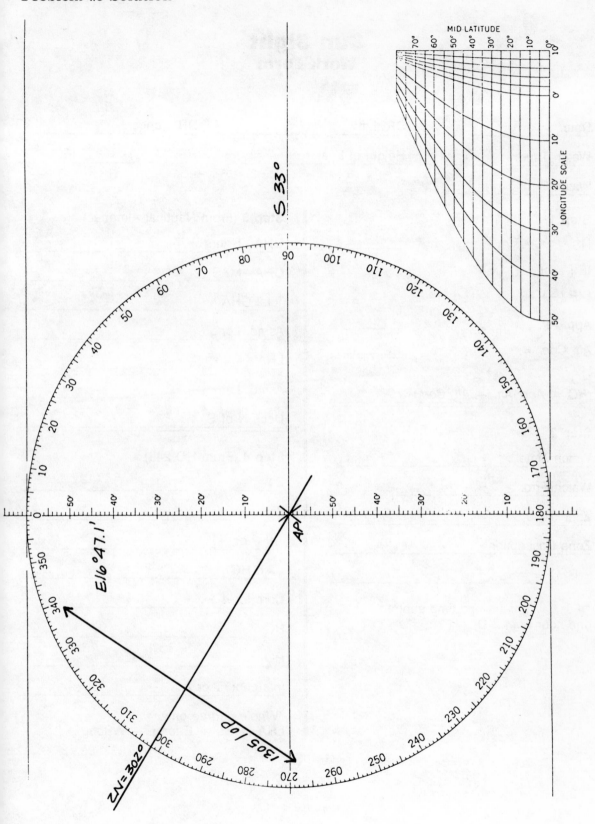

S 33°

MID LATITUDE

LONGITUDE SCALE

AP

E 116° 47.1'

1305 loP

ZN = 302°

111

Sun Sight
Work Form

Date: _____ DR. Lat: _____ DR. Long: _____

Watch Time: _____ Height of Eye _____ HS: _____ IE: _____

Watch Error _____

Step 1

HS _____

IE (+ or −) _____

DIP (−) _____

App. Alt. _____

Alt. Corr. = _____

HO =
(HO = App. Alt. ± Alt. Corr.)

Step 2

Watch Time   = _____

Watch Error ± = _____

Zone Time   = _____

Zone Description*= _____

  GMT = _____

*(For daylight-savings time subtract one hour from Z D)

Step 3 (From Nautical Almanac)

GHA—Hours _____

GHA—M&S _____

Total GHA _____

Asmd. Long. _____

LHA* _____

Asmd. Lat. _____

Decl. N or S _____

Step 4 (From HO 249)

Z _____

ZN _____

 d(+ or −) _____

 Tab HC _____

Corr. for d _____

HC _____

HO _____

Intercept, T or A _____

*Whole degree only
LHA = GHa + E long, − W long.

112

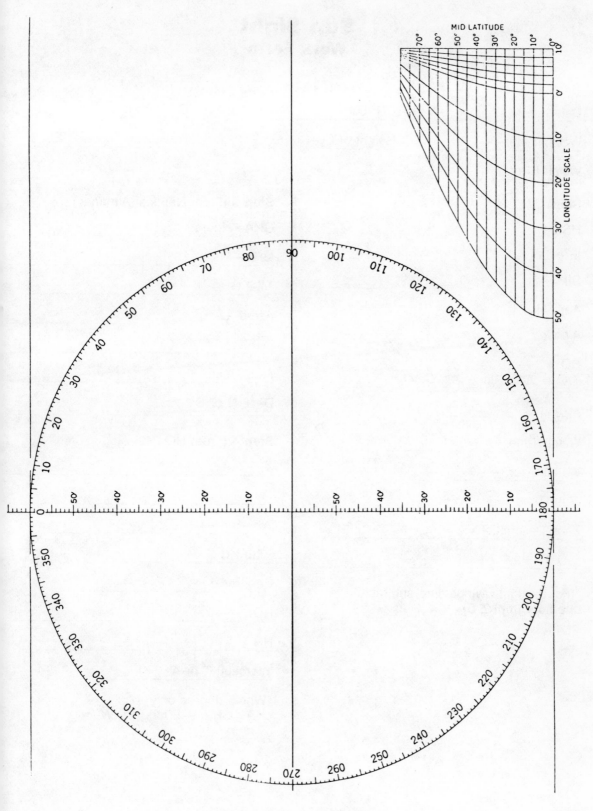

Universal Plotting Sheet

Sun Sight
Work Form

Date: _____   DR. Lat: _____   DR. Long: _____

Watch Time: _____   Height of Eye _____   HS: _____   IE: _____

Watch Error _____

Step 1

HS   _____

IE (+ or −)   _____

DIP (−)   _____

App. Alt.   _____

Alt. Corr. =   _____

HO =
(HO = App. Alt. ± Alt. Corr.)

Step 2

Watch Time   =   _____

Watch Error ±   =   _____

Zone Time   =   _____

Zone Description ≈   _____

  GMT =   _____

*(For daylight-savings time subtract
one hour from Z D)

Step 3 (From Nautical Almanac)

GHA—Hours   _____

GHA—M&S   _____

Total GHA   _____

Asmd. Long.   _____

LHA*   _____

Asmd. Lat.   _____

Decl. N or S   _____

Step 4 (From HO 249)

Z   _____

ZN   _____

 d(+ or −)   _____

 Tab HC   _____

Corr. for d   _____

HC   _____

HO   _____

Intercept, T or A   _____

*Whole degree only
LHA = GHa + E long, − W long.

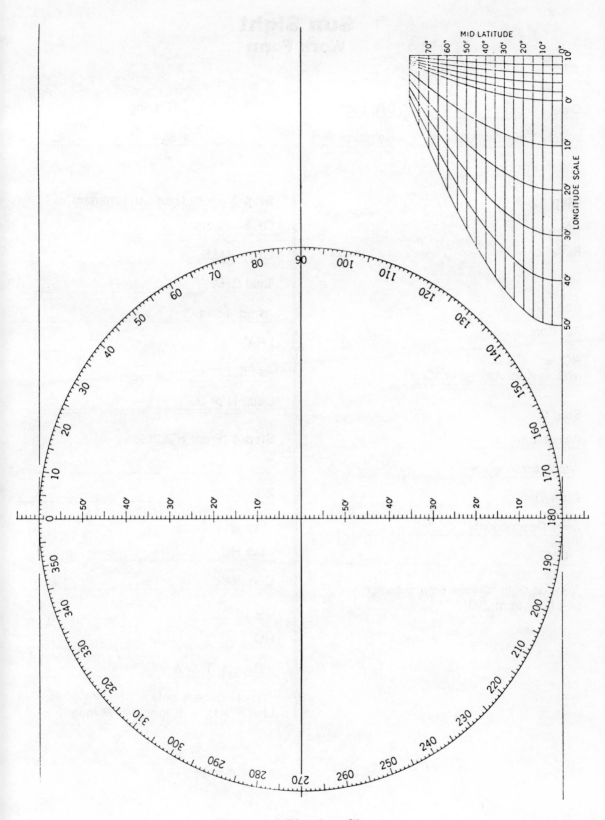

Universal Plotting Sheet

115

Sun Sight
Work Form

Date: _____

Watch Time: _____

Watch Error _____

DR. Lat: _____

Height of Eye _____

DR. Long: _____

HS: _____ IE: _____

Step 1

HS _____

IE (+ or −) _____

DIP (−) _____

App. Alt. _____

Alt. Corr. = _____

HO = _____
(HO = App. Alt. ± Alt. Corr.)

Step 2

Watch Time = _____

Watch Error ± = _____

Zone Time = _____

Zone Description = _____

  GMT = _____

*(For daylight-savings time subtract one hour from Z D)

Step 3 (From Nautical Almanac)

GHA—Hours _____

GHA—M&S _____

Total GHA _____

Asmd. Long. _____

LHA* _____

Asmd. Lat. _____

Decl. N or S _____

Step 4 (From HO 249)

Z _____

ZN _____

  d(+ or −) _____

  Tab HC _____

Corr. for d _____

HC _____

HO _____

Intercept, T or A _____

*Whole degree only
LHA = GHa + E long, − W long.

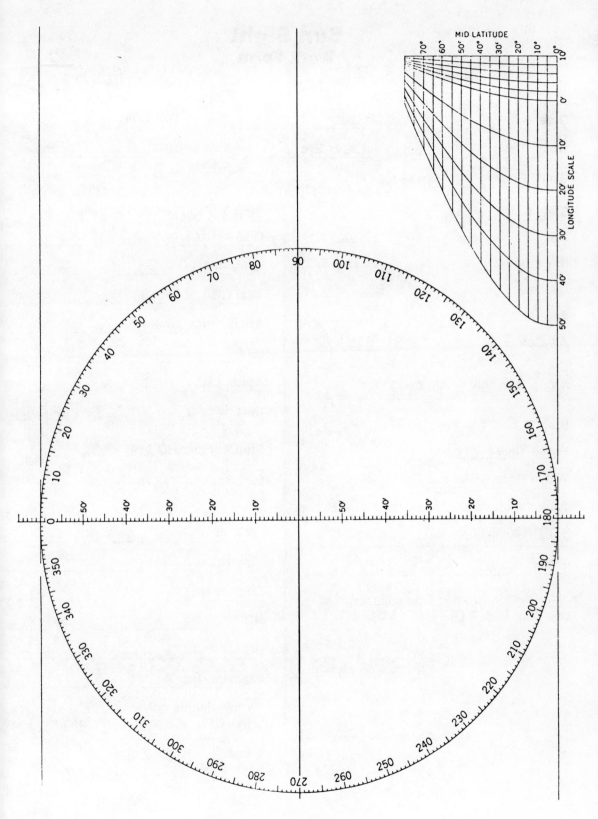

Universal Plotting Sheet

117

Sun Sight
Work Form

Date: _____   DR. Lat: _____   DR. Long: _____

Watch Time: _____   Height of Eye _____   HS: _____   IE: _____

Watch Error _____

Step 1

HS   _____

IE (+ or −)   _____

DIP (−)   _____

App. Alt.   _____

Alt. Corr. =   _____

HO =
(HO = App. Alt. ± Alt. Corr.)

Step 2

Watch Time   =   _____

Watch Error ±   =   _____

Zone Time   =   _____

Zone Description*=   _____

  GMT =   _____

*(For daylight-savings time subtract
one hour from Z D)

Step 3 (From Nautical Almanac)

GHA—Hours _____

GHA—M&S _____

Total GHA _____

Asmd. Long. _____

LHA* _____

Asmd. Lat. _____

Decl. N or S _____

Step 4 (From HO 249)

Z _____

ZN _____

 d(+ or −) _____

 Tab HC _____

Corr. for d _____

HC _____

HO _____

Intercept, T or A _____

*Whole degree only
LHA = GHa + E long, − W long.

118

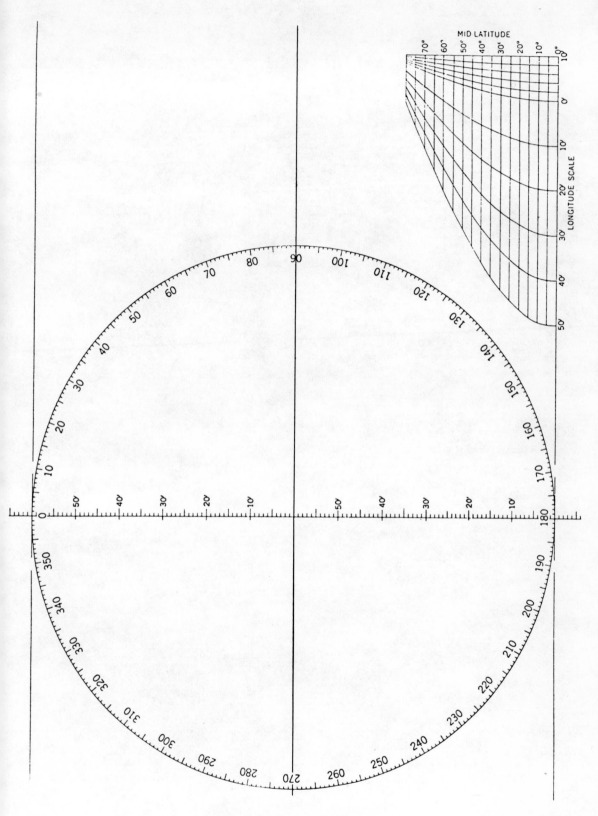

Universal Plotting Sheet

119